CONTENT

Chapter	Title	Page
1	Notes from Editors	4
2	The 96 Lections (aka Sections or Chapters)	44
3	Sample pages from Jain 108's publication from an early Manuscript	138
4	Comments on the 96 Lections	170

This manuscript is the 2nd edition of Jain's compilation, having a new cover based on the Essene Magic Square of 7x7, and computerized text.
Jain 108
2015, Mullumbimby Creek.

978-1-925834-33-8

nb: This Gospel record is the recovered document from which the Four Gospels as we have them today were built upon. It was the first formulated life of the Christ and was written by St. John about the year 70 AD., when he was imprisoned in Rome and given page by page to one whom he could trust.

When the scroll was completed and after its contents had been made known to the Apostles, it was taken to Tibet by the same disciple, who left it in the care of an unnamed lama. Here, it remained until a friar, named Placidus, visited the monastery in the eighteen seventies and asked if he might show it to the Church Authorities in Rome.

GOSPEL OF THE HOLY TWELVE

> ## The Gospel of the Holy Twelve
> *Translated from the original Aramaic*
> by Rev. Gideon Jasper Richard Ouseley M.A.

The Gospel of the Holy Twelve, also known as the Gospel of the Nazarenes, is the very Gospel that was repeatedly mentioned and described by many commentators of the early Church as the original teachings of the Nazarene called Christ.

---

taken from this website: 2015

http://www.thenazareneway.com/ght_table_of_contents.htm

CONTENTS

Cover
Foreword
Special Note
Explanatory Preface
To The Reader
Prologue

- Biography: Rev. Gideon Jasper Richard Ouseley M.A.
- Origins of the Gospel of the Holy Twelve
- The Legend of the Lost Gospel

A True Likeness of Our Saviour
Testimony of the Early Christian Church
The Epistle of Appollos the Prophet

The Comments of the Editors
Transcribed and Compiled by
Rev. Mark Wilcox, D.D.
Back Cover

Index of Lections 1 Thru 96

Section 1 Lections 1 Thru 10

(1) The Parentage and Conception of John the Baptist.
(2) The Immaculate Conception of Jesus the Christ.
(3) Nativity of John the Baptist.
(4) Nativity of Jesus the Christ.
(5) Manifestation of Jesus to the Magi
(6) Childhood and Youth of Jesus the Christ. He delivereth a lion from hunters.
(7) The Preaching of John the Baptist.
(8) The Baptism of Jesus-Maria the Christ.
(9) The Four Temptations.
(10) Joseph and Mary make a feast unto Iesus. Andrew and Peter find Jesus.

Section 2 Lection 11 Thru 20

(11) The Anointing by Mary Magdalene.
(12) The Marriage in Cana. The Healing of the Nobleman's Son.
(13) The First Sermon in the Synagogue of Nazareth.
(14) The Calling of Andrew and Peter. The Teaching of Cruelty in Animals.
(15) The Healing of the Leper and the Palsied. The Deaf man who denied that others could Hear.
(16) Calling of Matthew. Parable of the New Wine in Old Bottles.
(17) Jesus Sendeth Forth the Twelve and their Fellows.
(18) Jesus Sendeth Forth the Two and Seventy.
(19) Jesus Teacheth how to Pray - Error even in Prophets.
(20) The Return of the Two and Seventy.

Section 3 Lections 21 Thru 30

(21) Jesus Rebuketh Cruelty to a Horse. Condemneth the Service of Mammon.
(22) The Restoration of Iairus' Daughter.
(23) Jesus and the Samaritan Woman.
(24) Jesus Denounces Cruelty. Healeth the Sick.
(25) The Sermon on the Mount (part I)
(26) The Sermon on the Mount (part II)
(27) The Sermon on the Mount (part III)
(28) Jesus Releaseth the Rabbits and Pigeons.
(29) He Feedeth Five Thousand with Six Loaves and Seven Cluster of Grapes.
(30) The Bread of Life and the Living Vine.

Section 4, Lections 31 Thru 40

(31) The Bread of Life and the Living Vine. Jesus Teacheth the Thoughtless Driver.
(32) God the Food and Drink of All.
(33) By the Shedding of Blood of Others Is No Remission of Sins.
(34) Love of Jesus for All Creatures. His Care for a Cat.
(35) The Good Law. The Good Samaritan. Mary and Martha. On Divine Wisdom.
(36) The Woman Taken in Adultery. The Pharisee and the Publican.
(37) The Regeneration of the Soul.
(38) Jesus Condemneth the Ill Treatment of Animals.
(39) The Kingdom of Heaven (Seven Parables)
(40) Jesus Expoundeth the Inner Teaching to the Twelve.

Section 5, Lections 41 Thru 50

(41) Jesus setteth free the Caged Birds. The Bind Man who denied that Others Saw.
(42) Jesus Teaches Concerning Marriage. The Blessing of Children.
(43) Jesus Teaches Concerning the Riches of this World and the Washing of Hands and Unclean Meats.
(44) The Confession of the Twelve. Christ the True Rock.
(45) Seeking for Signs. The Unclean Spirit.
(46) The Transfiguration on the Mount, and the Giving of the Law.
(47) The Spirit Giveth Life. The Rich Man and the Beggar.
(48) Jesus Feedeth One Thousand with Five Melons. Healeth the withered Hand on the Sabbath Day. Rebuketh Hypocrisy.
(49) The True Temple of God.
(50) Christ the Light of the World.

Section 6 Lections 51 Thru 60

(51) The Truth Maketh Free.
(52) The Pre-existence of Christ.
(53) Jesus Healeth the Blind on the Sabbath. Jesus at the Pool of Siloam.
(54) The Examination of the Blind Man - A Living Type of the House of God.
(55) Christ the Good Shepherd.
(56) The Raising of Lazarus from his Sleep in the Tomb.
(57) Concerning Little Children. Forgiveness of Others. Parable of the Fishes.
(58) Divine Love to the Repentant.
(59) Jesus Forewarneth His Disciples. Glad Tidings to Zacchaeus.
(60) Jesus Rebuketh Hypocrisy.

Section 7 Lections 61 Thru 70

(61) Jesus Foretelleth the End.
(62) Parable of the Ten Virgins.
(63) Parable of the Talents.
(64) Jesus Teacheth in the Palm Circle. The Divine Life and Substance.
(65) The Last Anointing by Mary Magdalene. Neglect not the Present time.
(66) Jesus again Teacheth his Disciples concerning the Nature of God. The Kingdom of God. The Two in One.
(67) The Last Entry into Jerusalem. The Sheep and the Goats.
(68) The Householder and the Husbandmen. Order out of Disorder.
(69) The Christ within the Soul. The Resurrection and the Life. Salome's Question.
(70) Jesus Rebuketh Peter's Haste.

Section 8 Lections 71 Thru 80

(71) The Cleansing of the Temple.
(72) The Many Mansions in the One House.
(73) Christ The True Vine.
(74) Jesus Foretelleth Persecutions.
(75) The Last Paschal Supper.
(76) Washing of the Feet, The Eucharistic Oblation.
(77) The Agony in Gethsemane.
(78) The Betrayal.
(79) The Hebrew Trial before Caiaphas.
(80) The Sorrow and Penance of Iudas.

Section 9 Lections 81 Thru 89

(81) The Roman Trial before Pilate.
(82) The Crucifixion.
(83) The Burial of Jesus.
(84) The Resurrection of Jesus.
(85) Jesus appeareth to Two at Emmaus.
(86) Jesus appeareth in the Temple. Blood Sacrifices Cease.
(87) Jesus appeareth to the Twelve.
(88) The Eighth Day after the Resurrection.
(89) Jesus appeareth at the Sea of Galilee.

Section 10 Lections 90 Thru 96

(90) What is Truth?
(91) The Order of the Kingdom (part I.)
(92) The Order of the Kingdom (part II.)
(93) The Order of the Kingdom (part III.)
(94) The Order of the Kingdom (part IV.)
(95) The Ascension of Christ.
(96) The Pouring out of the Spirit. The taking of Mary and Joseph.

Gospel of the Holy Twelve
Front Cover

NEW EDITION

THE GOSPEL OF THE HOLY TWELVE

by

A Disciple of the Master

(Rev. G. J. R. Ouseley)

LONDON
JOHN M. WATKINS
21 CECIL COURT, CHARING CROSS RD., W.C.2

FOREWORD
SOME DEFINITIONS.

In theology or scientific religion, it is all important that terms should ever be used in one and the same sense, else confusion and error are and must be (as it has been) the result. Thus the word PERSON is used (in this work) in its primary signification *(persona)* appearance, or manifestation; while personality is that which belongs to *person.*

INDIVIDUAL (indivisible or undivided duality) is the One pure Spirit, the Eternal Esse, the LIFE and SUBSTANCE, the innermost, the all-pervading Spirit which resides in every person or manifestation.

Thus **GOD** being pure Spirit, the Innermost, the All- pervading, God's *Body* is the Universe in its totality, the ALL- PERSON or Manifestation of the hidden Deity. And each and every part thereof is a manifestation, and visible expression of Deity, according to its nature and degree. In this sense there is and can be but one All-person, All in all, and through all and around all.

HEAVEN is used for the Pleroma, Fullness of God.

HADES - The invisible - the place of departed souls, Purgatory.

GEHENNA denotes extinction, everlasting death (not eternal life in death).

AGE an existence or cycle of indefinite period.

AGES of ages *or* Greater Cycles consisting of such lesser cycles.

PERSONIFICATION, is the giving to An inanimate object the attributes of a living rational being; or again, the making of one Individual the representative or type of a whole race, or of a Messenger as a Manifestation of the Supreme. Thus Ioseph and Mary represented a Regenerate Humanity, and Iesu Maria, the Offspring thereof, first of the Sons and Daughters of God as a Unity.

Special Note

(From the first 1956 reprint edition)

As a relative of Dr. Anna Kingsford, that great lover of God's creatures, I am republishing this little book of "The Gospel of the Holy Twelve." I do so in the hope that its readers will be touched by the Compassion of the Master Christ for all living things, and that they will strive to imitate Him.

For the cruelties of the modern world towards animals and birds are a disgrace to civilized man. The agonies of the vivisection laboratories and slaughter-houses, of the victims of the fur trade, circuses and the rest, must be exposed in all their vileness, so that people everywhere may arouse themselves, and unite in ending them.

Let us remember that ALL LIFE IS ONE, and so guard from injury all living things, especially those most pathetically helpless and dependent upon us.

> "He prayeth best who loveth best
> All things both great and small;
> For the dear God who loveth us,
> He made and loveth all."

My thanks are due to Mr. S. H. Hart for his consent to this reprint. December, 1956

            RONALD HENTLAND

Gospel of the Holy Twelve
~ Explanatory Preface ~
To the New and Complete Edition
(Revised and Enlarged)

This "Gospel of the Holy Twelve" (Evangelists) of the Christian Dispensation is one of the most ancient and complete of early Christian fragments, preserved in one of the Monasteries of the Buddhist monks in Thibet, where it was hidden by some of the Essene community for safety from the hands of corrupters and now for the first time translated from the Aramaic. The contents clearly show it to be an early Essenian writing. This ancient community of the Jewish Church called Yessenes, Iessenes, Nazarites, or Nazirs, strongly resembling the Therapeutae, and the Buddhists, who practised community of goods, daily ablutions, daily worship, and renounced flesh eating, and strong drink and the sacrifice of animals, and the doctrine of "atonement" for the sins of some by the vicarious and involuntary suffering of others, as held by the Pharisees and Sadducees, and by the heathen before them; thus preparing the way for those Orders and Communities of men and women which have since arisen throughout the East and West, like cities set on hill, to shew the more perfect way to Christians living. in the world, notably those of S. Basil in the East, and S. Benedict in the West, and, with them, the Carthusians and the Franciscans, and before them all, the Carmelites (who had their headquarters on Mount Carmel) to whom they are similar in their customs, and even their dress, if not altogether identical with them, tracing their origin to Elias, abstaining from all flesh meats and strong drinks, whose symbol was, it is said, an iron cross in a circle, and among the animals, the Lamb and the Dove their special emblems. See Philo (in Loco) or Kitto's Cyclopaedia (art, Essenes), also Arthur Lillie's "Christianity and Buddhism."

That the contents of this most ancient Gospel set forth a higher moral and religious teaching, as the basis of the Christian Church, than any other that has come down to us, requires but the reading of eyes divested of prejudice, and the perception of a regenerate heart, and intelligent mind, to receive and appreciate. The giving of the New Law on the Holy Mount is a scene that, once read, can never be forgotten, though it was not "with blackness and thunder and the sound of the trump."

Inasmuch as this Gospel touches on many questions of vital moment now discussed in this age, and little known in those times, it may well be termed par excellence the prophetic and ethical Gospel, and critics and scholars will remember that the writings of Justin Martyn, Papias and others, distinctly speak of, and quote from, the "Gospel of the Hebrews " known otherwise as "the Gospel of the Twelve Apostles " and the "Gospel of the Nazarites," used then, chiefly in the Church at Jerusalem, and the original of Matthew's Gospel in Hebrew which we have in Greek. This identifies it as the original Gospel from which the others were more or less closely copied, with numerous variations and important omissions by accident, or design, to suit the corrupt taste of the worldly.

To those who may think it a difficulty, that the name of this notorious body (the Essenes) not less notorious than the Pharisees or Sadducees, so often spoken of, is not once mentioned either in this Gospel, or in the received four, the very silence ought to appeal, as an eloquent testimony to its origin.

It is the opinion, after much patient study and research of that learned traveller and author, Mr. Arthur Lillie, in his works on "Buddhism and Christianity " that "the earliest and only authentic Gospel must come from the Essenes, and all that is anti-Essene (in our four) is accretion."

This new and complete edition we have revised and enlarged, not only in the explanatory preface (itself enlarged from the second) but also by additional matter and corrections which escaped our notice in first issue, and also important notes, which may prevent misunderstanding by the unthinking, and the wrong thinking, who form the vast majority of readers in this age. These are now incorporated with this edition. The number of verses in each lecture have been better arranged, while the number of Lections remain the same. The writers have been at great pains to verify in the early Christian writers most of these older utterances of Iesus, strange in modern ears, as far as their present limitation would admit, many of them being utterly lost to the world till the present revelation.

Of Placidus (cir. 1326), he was known only by the Editor as that Carmelite monk, sitting occasionally by his bedside and discussing many questions. To the query, "Suppose that in these days of discovery of ancient documents some ancient and genuine MSS. were found in which it was clearly written, 'Joseph begat Jesus of Mary his wife, who was afterwards called the Christ,' Do you think the Church dogma of the immaculate conception would be imperilled to your mind ? " he instantly replied, "Not in the least, the Church would in that case propound and define the immaculate conception of Joseph as well as his bride - the safeguard would be doubled." Judge of the Editor's surprise, when within six months after news came from the East of the discovery by Mrs. Lewis of a most ancient and undoubtedly genuine MSS. of the Holy Gospels, with these very words in the first chapter of St. Matthew, preceding the interpolation, without which the sense of the entire is evident and complete.

Of John Wesley (d. 1791) also it may be said that he, too, was of the number that ate not flesh nor drank strong drink. Writing to the Bishop of London in 1747 , he said: " Thanks be to God, since the time that I gave up the use of flesh and wine, I have been delivered from all physical ills."
As formerly, the writers and the transmitters of sacred Scriptures (in their purity for the benefit of humanity) were sometimes forced, by motive of prudence, to conceal their personalities while in the flesh, knowing well, that for a divine message, its own truth was the best evidence of genuine inspiration, and that by reason of its truthfully rebuking the cherished views of the many, they would meet with the usual reward from the ignorant and worldly-persecution of one kind or another - misrepresentation of motives, personal depreciation (the favorite refuge of those who cannot refute, and will not accept)- so now for the same reason, seeking neither praise nor blame from men, the visible Editors of these Scriptures withhold their names. For the world is the same now as then, yea, worse, for these are the days of unbelief in God, the Good, the Beautiful and the True- "the rude latter days" of which the late Father Faber spoke so touchingly - when courtesy is almost unknown, and reverence and faith are well-nigh dead, and obedience to God is counted nought, when it would restrain or interfere with some perverse and wicked fashion or folly of the day, or some deep-rooted evil habits or customs of long standing.

As this Gospel was not addressed to the heathen, but chiefly to the true followers of Iesus, in the early days of the Church of Jerusalem, so now it is sent to modern Christians who have fallen into worse than heathen darkness, if perchance it may be received by a few men and women of "Peace and Goodwill" to whom "Peace on Earth" was originally announced. It is quite immaterial to the Editors whether it be or be not received, though to them who reject or ignore it, it may be otherwise.

For long ages, since the destruction of Jerusalem, the aspect of God, as the Eternal Mother ("Holy Spirit"), One with the Eternal Father, has been concealed from our eyes by clouds of human tradition, error and superstition, but now the Sun of Righteousness (the Divine Mother}, shines in its fullness from behind the clouds of darkness that have so long hidden it from view, and it shines alike on believer and unbeliever, on those who see and feel its warmth, and on those who will to remain in darkness, and perish for want of light. It is written, "Behold I come from behind the clouds and every eye shall see Me whom they have pierced, and all of the earth who see, shall wail because of their iniquities. "The world is deaf to the beloved Voice, and hears not, blind and sees not, but only the things which make for its own lusts. As of yore, approaching to the Holy City, Jesus wept, so now to modem Christendom He seems to say, "Ierusalem, Ierusalem, thou that stonest the prophets and revilest them that are sent unto thee, how often would I have gathered thee as a hen gathereth her chicken under her wings, but ye would not! Oh that thou hadst listened in this thy day to the things which make (through righteousness) for thy peace, but ye would not! Behold the day cometh, when thine enemies shall cast a trench before thee and surround thee on every side, and burn thee with fire, leaving not one stone upon another. Behold now is your house left unto you desolate, and ye shall not see Me henceforth, till ye shall say, Blessed are They who come in the Name of the Holy One."

THE EDITORS OF THE GOSPEL OF THE HOLY TWELVE.

Gospel of the Holy Twelve
TO THE READER

THE all-pitying love of our Saviour embraces not only embraces mankind, but also the "lower" creatures of God, sharers with us of the one breath of life, and with us on the one road of ascent to that which is higher. Never has the Providence with which the All-Merciful watches over the animals "unendowed with reason," as well as over "reason-endowed" man, been more impressively brought home to us than in the saying of Iesus : "Are not five sparrows sold for two farthings, and not one of them is forgotten before God ?"

How were it possible to think otherwise, than that the Saviour "would have had pity and compassion on the creatures who must bear their pain in silence. Would it not seem to us like a blasphemy if we were to hear it said that Jesus or Mary would have beheld, without pity or succour the ill-treatment of helpless animals? Nay, certainly, when our Saviour brought redemption to a world sunk in selfishness, hard-heartedness and misery, and proclaimed the Gospel of an all-embracing love, there was, surely, a share in this redemption for all suffering creatures, since when man opened his heart to this Divine Love, there was no room left for pitiless hardness towards the other creatures of God, who have, like man, been called into life with the capacity of enjoyment and of suffering. They bear the marks of the Redeemer, who practises this all-pitying love. And how little it is that the minimum of Christian compassion for helpless creatures demands of us! Only to inflict on them no torture, and to help them when in trouble, or appeal to us for succour, and when, of necessity, we take their life, to let it be a speedy death with the least pain; a gentle sleep. But, alas! how little are we penetrated with these divine lessons of mercy and compassion. How many grievous tortures are inflicted on helpless creatures under the pretence of science, or to gratify an unnatural appetite, or cruel lusts, or the promptings of vanity.

As an aid to a higher Christianity these fragments of a fuller Gospel are now presented, giving us the Feminine tenderness as the Masculine strength of the Perfect Christ.

The greater and more important portion of these reminiscences have formed the groundwork and basis of various teachings issued by the Order of At-one-ment since 1881, when it was incepted, and are now for the first time given in their entirety, throwing additional light on the real doctrine of Iesus, or elucidating of the contents of the canonical Gospels as commonly received, retaining the translation of the A.V. wherever possible, or sufficiently clear. It will be for the Church of the Future when revising the, entire Scriptures to give it its primary place, the original and complete "Gospel of the Holy Christ," using the others for a confirmation from four other witnesses that every word may be established to them who are not in a condition to receive the goodness, purity and truth of the former.

Like all other inspired writings (but not necessarily infallible in every word) these writings from within the Veil must be taken on their own internal evidence of a Higher Teaching. For inspiration of the Spirit no more necessarily implies infallibility than the divine breath of life inbreathed by man, necessarily implies freedom from all accidents, diseases or miseries incidental to mortal life.

It is a faithless and perverse generation, as of old, that seeks for signs, and to them saith the Spirit, 'there shall no sign be given," for were the very writers of this Gospel raised from the dead, and were they to testify to their authorship, they would not believe, unbelieving critics would still ask for a sign, and the more they were given the more they would ask in the hardness of their hearts. The sign is The Truth-the pure in heart they shall see it.

The Gospel of the Holy Twelve
Translated from the original Aramaic
by Rev. G.J.R. Ouseley

**IN THE NAME OF
THE ALL HOLY. AMUN.**

Here beginneth the Gospel of the Perfect Life of Jesu-Maria, the Christ, the offspring of David through Joseph and Mary after the flesh, and the Son of God, through Divine Love and Wisdom, after the Spirit.

PROLOGUE

From the Ages of Ages is the Eternal Thought, and the Thought is the Word, and the Word is the Act, and these Three are one in the Eternal Law, and the Law is with God and the Law proceeds from God. All things are created by Law and without it is not anything created that existeth. In the Word is Life and Substance, the Fire and the Light. The Love and the Wisdom, are One for the Salvation of all. And the Light shineth in darkness and the darkness concealeth it not. The Word is the one Life-giving Fire, which shining into the world becometh the fire and light of every soul that entereth into the world. I am in the world, and the world is in Me, and the world knoweth it not. I come to my own House, and my friends receive Me not. But as many as receive and obey, to them is given the power to become the Sons and Daughters of God, even to them who believe in the Holy Name, who are born--not of the will of the blood and flesh, but of God. And the Word is incarnate and dwelleth among us, whose Glory we beheld, full of Grace. Behold the Goodness, and the Truth and the Beauty of God!

The Nazarene Way of Essenic Studies
Biography: Rev. Gideon Jasper Richard Ouseley M.A.
1835 - 1906
Translator of the Gospel of the Holy Twelve.

The Rev. Gideon Jasper Richard Ouseley was born in Lisbon on the 15th October, 1835, the younger son of Sir Ralph Ouseley, K.C.B., brought over to Ireland on the death of his father, 14th May, 1842, by relatives, and educated in Dublin University, in which he graduated in 1858. On the 9th of December, 1906 he passed away in his seventy-second year.

He was ordained as a clergyman of the Established Church by the Bishop of Down and Connor, and appointed Curate of Warren- point, Co. Down in 1861.

He did not long remain in a Church in which he found himself to be in a false position; and in 1870, having voluntarily renounced all eating of flesh, strong drink and tobacco as inconsistent with the humanity and the true religion of Christ, as taught by Him and His apostles, he was received as a priest of the Catholic Apostolic Church.

Before his reception into the Catholic Church, he fully explained to the Prior his views and beliefs—which on no account would he abandon. The Prior said he did not find in that which he expressed anything contrary to the faith of the Catholic Church. "Why then," asked the Rev. Ouseley, "does not the Church teach it?" To which the Prior replied, "You are not at liberty to say it is the teaching of the Catholic Church, but you are free to believe and hold your views as a pious opinion."

It was in such circumstances that he sought to propagate his beliefs. His desire was to saturate and transmute the Church's own formularies with the true teaching of Christ and His Humanitarianism. He set about revising the formularies in place of concealing them.

While he was not a sectarian and hated to be dubbed as such, he would work with all true reformers and good men who would try to bring the people into ways of righteousness and raise them out of their barbarous habits of flesh-eating, strong drinking, and cruelty of every sort and kind.

He considered that "the direct cause of poverty, bad health and social misery was due to flesh-eating, alcoholic drinking and tobacco smoking". He saw in their abolition "the only effectual means of the world's redemption, whether as regards men themselves or the animals". "The true and proper food for man was that which Mother Earth brought forth in plenty for the sustenance of her children".

He founded the Order of At-one-ment and United Templars Society, having for its motto: "One God, one Religion, various names, various forms." Its object was to bring about a reconciliation of opposing ideas, things, persons, and systems; at-one-ing the human with the Divine, and man with God; by the Spirit of Christ within the soul.

On account of his views, the personal relationship between the reformer and his Church became strained; which led, in the first instance, to the Rev. Ouseley being suspended, and later to his being, literally, cast out of the Church. His views being called "Anti-Christian."

This because, believing that he had been called to undertake this important mission, he set about to present to the world the hidden text "The Gospel of the Holy Twelve," the Christian Dispensation pieced together from the most ancient and complete collection of Christian fragments, preserved in one of the Monasteries of the Buddhist monks in Thibet, where it was hidden by some of the Essene community for safety from the hands of the Corrupters and for the first time to be translated from the Aramaic. It was a translation of an original Aramaic document purporting to be a reconstruction and revision of the Gospel narrative.

Also because the contents of this most ancient Gospel set forth a higher moral and religious teaching, as the basis of the Christian Church, than any other that has come down to us, requires but the reading of eyes divested of prejudice, and the perception of a regenerate heart, and intelligent mind, to receive and appreciate that the Rev. Ouseley was undaunted in proclaiming the true religion of Christ.

One thing is unquestionable, he could not unaided by some Power higher than and above that of his normal intellect, have written this Gospel, and that such Power was of a Divine nature, is manifest from its contents. At the end of the Gospel are the words: "Glory be to God by Whose Power and help it has been written."

It must not on this account be assumed that the original of which this Gospel is claimed to be a translation is of greater historical value than are the Four Gospels— none of which is historical in the sense commonly supposed—the object of all being to portray "the divine drama of the Spiritual history of man"—that is, of the Soul. But this Gospel has the advantage over the others in that it escaped the hands of the "Correctors" and "Corrupters" from which they suffered.

In 1905, he wrote to a friend, "I have finished my work and my destiny awaits me. They are ready to receive me".

On the 9th of December, 1906 he passed away in his seventy-second year. A notice of him appeared in "Light" (5th and 12th January, 1907): "And now may he rest in peace with the knowledge that his life of self-sacrifice for Truth and Humanity—a life without ambition for any worldly honour or desire for any worldly gain—was not lived in vain, for "Where Mercy, Love and Pity dwell, There God is dwelling too."

The Nazarene Way of Essenic Studies

Origins of the Gospel of the Holy Twelve

The Original Gospel Record Revealed

This Gospel record is the recovered document from which the Four Gospels as we have them today were built upon. It was the first formulated life of the Christ and was written by St. John about the year 70 AD., when he was imprisoned in Rome and given page by page to one whom he could trust.

When the scroll was completed and after its contents had been made known to the Apostles, it was taken to Tibet by the same disciple, who left it in the care of an unnamed lama. Here, it remained until a friar, named Placidus, visited the monastery in the eighteen seventies and asked if he might show it to the Church Authorities in Rome.

The scroll was then given to him and on the way home, which took a long time, he translated a portion of it into Latin and on his arrival in Rome read it to the assembled Cardinals. At first they were impressed, but as he proceeded to reveal the contents they realised that to make it known would discredit the Church which had, during the Council of Nicea, eliminated from the Gospels the Master's teaching about the love and care of animals and about abstaining from eating flesh-foods. And so the scroll was hidden away in the archives of the Vatican, where it remains to this day.

The reason this Gospel contains much that is in St. Matthew, St. Mark, St. Luke and St. John is because there were many Gospels written about this time containing the true teaching of Jesus, but this was the only one that escaped the pen of the `correctors', because they did not know of its existence.

The other Gospels contained the teaching of Jesus about the avoidance of meat-eating and about the love for animals, but all this teaching was eliminated by the `correctors'. "The Gospel of the Holy Twelve" is an authentic Gospel and should be accepted just as it is as the original teaching of Jesus.

Like all Sacred Scriptures, The Gospel of the Holy Twelve is mystical, and for right understanding it must be so interpreted. Its value lies in the teaching which in the garment is veiled. To understand the teaching, the garment must be lifted or removed. Then "falsehood passes away, but truth remains."

In all sacred mysteries, parables are used as garments for Truth that is hidden in its very expression. In one of the `Sayings of Jesus' —as recorded in the *oxyrhynchus Papyri—we are told: "That which is hidden from thee shall be revealed to thee. For there is nothing hidden which shall not be made manifest, nor buried which shall not be raised." To those who hid "the key to knowledge" Jesus said "Ye entered not in yourselves and to them that were entering in ye opened not." Ultimately, "Truth itself is unutterable save by God to God."

As stated in the Gospel of the Holy Twelve: "Truth is absolute and is in God alone. To men is Truth revealed, according to their capacity to understand and receive." "Look for the sufficient meaning of the manifest Universe and of the written word, and thou shalt find only the Mystical Sense."

The Gospel Of The Holy 12

"To some I speak things common. To others in great light I reveal mysteries: My words are Spirit and Life and are not to be estimated by the sense of Man. They, the prophets, deliver the letter but thou discloseth the sense. They publish mysteries, but thou explainest the meaning of things sealed. They cry with words, but thou givest understanding to the hearer." At the close of the Gospel of the Holy Twelve we read: "For them that believe regarding the spirit rather than the letter which killeth, the things herein related are true as Spiritual verities, for others they are as an idle tale."

It was not without profound meaning that Jesus said: "Raise the stone" of the indwelling Self (to the level of Spirit), and there thou shalt find Me": and "Cleave the wood", of the lower and outer consciousness, "there am I". The "Son of God in man" must be lifted up for the right understanding of the Holy Scripture. Jesus said unto "the sick of the palsy"—and there are many such—"Arise, and take up thy couch."

One of the first acts of Jesus after His resurrection was to "open the Scriptures" to two of his disciples—-Cleophas and another— whom He joined on their way from Jerusalem to Emmaus. These disciples He found "reasoning together" about the things which had then lately come to pass; and forthwith "He expounded unto them in all the Scriptures, the things concerning Himself"; and He blamed them for their "folly" and for their "slowness of heart to believe all that the prophets have spoken."

Had they listened to the dictates of their hearts rather than to the reasoning of their minds, their Spiritual eyes would not have been "holden." Jesus also said: "My soul grieveth over the sons of men, because they are blind in their heart and see not," and again Jesus warned his followers to "beware of false prophets which come in sheep's clothing, but inwardly are ravening wolves." They would be known "by their fruits."

The fact that the disciples reasoned together implies a limitation of their consciousness to the outer and lower mental planes, wherein are seen but "the shadows of the tomb." Thus were the Spiritual eyes "holden". Their "Heart" was "slow"—a slowness which indicated lack of spiritual understanding. But "While He talked with them by the way, and while He opened to them the Scriptures", a change came over them: "Their hearts did burn within them", and their Spiritual eyes were opened.

They awoke and knew Him. He was known of them in the "breaking of the Bread" of Divine Truth which is the food of the soul. And on the opening of their Spiritual eyes, He whom therefore they had known outwardly. "Vanished from their sight", and they were enabled to testify: "The Lord is risen indeed"!

Most of the mistakes of the materialists arise from understanding localities and things material, when they should understand conditions and principles. The letter of Scripture is but the shadow of Divine Truth.

Those who mistake the shadow for the substance and the substance for the shadow, never arrive at Reality but follow false aims. Those who know the substance as the substance and the shadow as the shadow, arrive at Reality, and follow right aims.

Why are certain teachings hidden beneath the letter? Why, it may be asked, did Jesus teach the multitude in parables? The answer is, the Mysteries cannot be taught by the blood-guilty, nor may they except as "dark sayings", be given to those—the multitude—who live in conflict with Christ's teachings.

On this the Gospel of the Holy Twelve is explicit: "They who partake of benefits which are gotten by wronging God's creatures, cannot be righteous", nor can they "whose hands are stained with blood, or whose mouths are defiled with flesh", understand the mysteries of the kingdom, nor are they fit to receive the highest mysteries—"For this people's hearts are waxed gross".

The Aramaic "fragments" to which reference has already been made also deal with the "healing works of Jesus", many of which were effected by methods now known as nature-cure. It is therein recorded that Jesus said: "He who kills, kills himself; and whoso eats the flesh of slain beasts, eats of the body of death. For in his body every drop of their blood turns to poison, and their death will become his death." The command against killing is insisted on throughout, and the eating of all dead flesh is condemned.

The age in which we live is no age of faith. It is materialistic and anti-christian. Wickedness has become legalised. There is materialism in science and materialism in the Church and in religion. Truth is being suppressed and people are losing faith in so- called democracy and reformers ignore religion and Religion is therefore, and no wonder, repellent to many in these days.

It is important to bear in mind that while the souls of the Righteous become vehicles of Divine manifestation, some inspired writings are coloured by the character or mentality of the instrument or medium through whom they come.

They are not always to be regarded as infallible in every word. Inspiration of the Spirit does not necessarily imply infallibility of utterance, because no man is wholly without error. In the Gospel of the Holy Twelve we read that even among the prophets there has been found "the word of error".

For a divine message its own truth is the best evidence of genuine inspiration; which is the inbreathing of the Divine through the spiritual organs planted in man for that purpose. Inspiration has its birth in God. To men truth is revealed according to their capacity to understand and receive it.

The creation and salvation of the world is achieved "By the Descent of Spirit into Matter, and the Ascent of Matter into Spirit through the Ages," and such is the teaching of the Gospel of the Holy Twelve. We also find in this Gospel a command of Jesus that we "Love one another and all the creatures of God"—God being in all creatures.

Children are to be "brought up in the ways of righteousness, neither eating flesh, nor drinking strong drink; nor hunting the creatures which God hath given into the hands of man to protect." That Jesus came into the world to put an end to bloody sacrifices and flesh eating is explicit.

The doctrine of Regeneration is definitely taught by Jesus as the means whereby man is "made perfect through suffering: changes of life for the perfecting of souls." Having thus become "purified through many experiences," man shall die no more, neither shall he be born any more, for death hath no more dominion over him. In this teaching, there is no "scapegoat" Christianity: nor forgiveness or remission of sin by "vicarious atonement": For sins against the law of God there can be no remission save by "repentance and amendment."

"The Scriptures contain the Word of God, but often interpolated and transformed by the error of man, whether by accident or design.* Shall we cast away the gold or despise it for the sake of the dross mingled with it? Doing this we should be foolish, not wise."

The purpose of making known the Gospel of the Holy Twelve is for the restoration of the original text, depicting the life and uttering the truth relating to Christ's mission on earth. In this Gospel the prediction by Jesus was fulfilled. He said to His disciples: "They shall put you out of the Synagogues because they have not known me". And if it be asked who are His disciples, the answer is given thus: "By this shall all men know that ye are my disciples if ye have love one to another, and shew mercy and love to all creatures of God."

Let us bear in mind the following words recorded in the Oxyrhynehus Papyri: "Jesus saith, let not him who seeks cease until he finds, and when he finds he shall be astonished; astonished he shall reach the Kingdom, and having reached the Kingdom he will rest."

Let it be said that no mere ecelasiastical organisation will in the long run be able to stand against the eternal principles of Justice, Humanity and Love to all manifestations of God, whether human on non-human.

Of the Churches it may be said—"We sat on the ancient foundations, but we revealed not the ancient truth; we have the keys to heaven, but we opened not the gate ourselves nor suffered others who desired, to enter. To us was given Light, but we concealed it in a dark place, and those who cried for more light we persecuted and counted as heretics, and caused many to be put to death in our blindness.

And even now. 0 Master, we had well nigh again rejected Thee, but by the mercy of the Eternal—we heard the Holy and True doctrines which once Thou gayest by parable—even the doctrines of ancient times given anew—the new wine of Thy Kingdom; and at length our eyes being opened and our ears being unstopped, we have returned to Thee."

As this Gospel was not addressed to the heathen, but chiefly to the true followers of Jesus, in the early days of the Church of Jerusalem, so now it is sent to modern Christians who have strayed into worse than heathen darkness; and if perchance it may be received by men and women of "Peace and Goodwill" to whom "Peace on Earth" was originally announced, the effort in sending it forth will be amply rewarded.

THE CHRISTIAN GOSPEL TRUST,
HENLEY BRIDGE HOUSE,
ASI-IBURiNHAM, BATTLE,
SUSSEX.

Notes:

* Oxyrhynchus Papyri: the Collection of many thousand fragments of papyri, found in 1897 onwards at Oxyrhynhus, a centre of Christian culture in the 4th Century, some 10 miles west of the Nile, near the modern Behnesa. The most celebrated are two series of "sayings of Jesus".

* Archdeacon Wilberforce of Westminster said "that after the Council of Nicea, AD 325, the MSS. of the New Testament were considerably tampered with". And Professor Nestlé in his "introduction to the Textual Criticism of the Greek Testament" tells us that certain scholars, called correctors, were appointed by the ecclesiastical authorities, and actually commissioned to correct the text of Scripture in the interest of what was considered ORTHODOXY.

"Beyond question of doubt", says Manley Palmer Hall, "records concerning Jesus do exist. It is equally certain that they are in the hands of people (the Vatican in Rome) who do not intend to make them available as they would endanger the institution of Christian Theology".

The Gospel of the Holy Twelve
The Legend of the Lost Gospel

Continuing archeological discoveries this century are shedding ever greater light onto the most important and mysterious period in human history; the man and the message of Jesus the Nazarene. Given the recent discoveries of the "Dead Sea Scrolls", the "Gnostic Gospels" of Nag Hammadi, and now, the long-sought "Gospel of the Nazirenes", far more has been discovered about the earliest days of Christianity in this century than in all the years prior. And what is becoming more and more clear from these discoveries is that the original message of Christ differs sharply from that of the official doctrines later adopted and enforced by the church.

The Discovery of Christ's Original Teachings

In 1870, an Aramaic manuscript entitled *"The Gospel of the Nazirenes"* was discovered, translated and published. This ancient scripture, hidden away for centuries in a Tibetan monastery, seems in virtually every respect identical to the work by the same title, that was known and widely quoted from in the first century by the church. Many of the most revered early church fathers, as well as a surprising number of scholars today, have boldly declared that the legendary *Gospel of the Nazirenes*, later to be known as *"The Gospel of the Holy Twelve,"* is nothing less than the long-lost original Gospel which, legend holds, was collectively written by the actual 12 apostles in the period immediately following Christ's death, and upon which all of the Biblical synoptic Gospels are based.

The Legend of the Lost Gospel

For nearly 2,000 years, all we objectively knew of Jesus came to us primarily through the Biblical Gospels. And yet, for all this time, a great and enduring enigma has loomed over these lofty works. In the fourth century, the ruling authorities of Rome decided which of the countless texts, based on Christ's teachings in circulation at that time, would make up the present-day Bible and deciding once and for all, in effect, which works were to be judged as authoritative and which were not. This decision, unfortunately, carried the undeniable taint of political compromise, and the Bishops making these decisions were doing so at the direct command of the Roman Emperor, and their future financial and social well-being was, and everyone agrees, entirely under his control. It has been whispered ever since the fourth century that much of the true message of Jesus was edited out at that time, due to the oppressive and theologically obtuse influence of Constantine.

The Christian scriptures that failed to be admitted into the Bible were then outlawed, collected, and destroyed.

Prior to 325 AD, however, many of the early Church fathers had included in their writings mention of an earlier Gospel, upon which they claimed in near-perfect unison, the synoptic Gospels of Matthew, Mark, and Luke had all been based. Mentioned or quoted from by such well-known church fathers as Papias, Hegesippus, Iranaeas, Clement, Origen, Basil, Epiphanius, Eusebius, and St. Jerome, this document had gone variously by the title *"Gospel of the Nazirenes"* (The word "Nazirene" comes from the "Nazirite-Essene" sect, or a Nazirite sect of the Essene branch of Judaism), *"Gospel of the Hebrews", "Gospel of the Ebionites"*, and *"The Aramaic Gospel of Matthew"*.

For nearly 2,000 years, historians considered this work to have been irrevocably lost, but in 1870 a forgotten copy was discovered, hidden away in a Tibetan monastery, and was quickly translated from the original Aramaic, published this time as *"The Gospel of the Holy Twelve"*. This work was translated into the old-style King James English by Rev. G.J.R. Ouseley. The work was quickly rejected however, and considered blasphemous by the Church.

This text certainly appears to be the very same gospel referred to by so many ancient commentators. Although this original scripture had indeed thought been lost, a number of its passages are well known, having been preserved by various church fathers who quoted them in their writings. Virtually all of the quoted passages can be found in this original Gospel and in their entirety. (as are also virtually the entire contents of the Gospels of Matthew, Mark, and Luke).

Numerous historical references thus seem to confirm the authenticity of the 1870 manuscript, and many modern scholars since 1870 have concluded as well that this work is, in all likelihood, the original source of much of the material that eventually found its way into the Biblical Gospels. If so, the Biblical Gospels would then be mere edited versions of this earlier, and therefore more authoritative work, just as many have argued over the centuries.

The Stamp of Authenticity

Far more than that of the Biblical Gospels, this work has the feel of having been written by actual witnesses to the events it describes. The detail is often both natural and more explicit, and a great many theological, social, and political issues come out making a great deal more sense.

Often during the reading of this work, one feels that you are simply reading the Bible, for many passages are, indeed, virtually identical to that found in the canon. The familiar old stories are told again, and either the wording is identical, or, when expanded upon or alternate wording is used, the stories come out making rather more sense than before, clearing up many questions left hanging in the "authorized" version. Never does it seem that the material is out-of-place, or as if it had been pasted-in by editors after the fact. Rather, in virtually every instance the fresh material seems an integral component of the narrative as one reflects anew upon the familiar wording of the "authorized" Bible.

Enriching details run throughout this text, giving the compelling impression that this is indeed an original eyewitness narrative, not a bland, confused, or glossed-over retelling of a dusty tradition repeatedly handed down orally for 30 years or more before finally being committed to writing. Traditional scriptural teachings maintain that the mighty works recorded in the New Testament went unwritten for some 30 years or more before being put down in writing, but this seems unlikely. At least some of the apostles were, reportedly, quite literate and learned men, and it seems, even prior to encountering a text such as this, that an already close-knit group of 12 accomplished apostles of Christ would have quickly pooled and compared their memories in an effort to compose a definitive version of their recollections of the man, his teachings, and the works of Jesus, before anything of importance could be lost or forgotten.

This sacred text, now here available for all to read, constitutes evidence that such a collective testimony not only was composed (just as reason suggests it would have been), but has successfully survived the centuries after all, even in spite of whatever political forces that might at one time have been aligned against it.

It seems as if the authorized Gospels in the present day Bible are all various edited versions of the *"Gospel of the Holy Twelve"*. Some material originating in this text has even found its way into the biblical books of Acts and Revelations.

In many ways, while reading it, the familiar age-old message of the Bible comes through as always; but then one is suddenly jolted upright, reading startling passages that directly defend the very non-Western traditions of reincarnation, the female aspect of creation, and compassion for *all* creatures along with the equally unfamiliar tales of Jesus' studying various mysteries and wisdom traditions in India, Persia and Egypt.

In many places, then, what is written in this text contrasts sharply with the familiar story and message in the "authorized" Bible. It teaches strict and uncompromising vegetarianism, describing how Jesus' anger at the Temple was not merely directed at the financial business going on there, but was *specifically* over the selling and slaughtering of sacrificial animals in the Temple, which was supposed to be a House of Prayer, but had been changed, he cried, "into a slaughterhouse." The idea that Jesus might have felt outrage at seeing the cruel carnage of innocent creatures in the Holy Temple seems fully consistent with his character as we have collectively come to imagine Him, and this interesting variation of the "moneychangers" event comes across as a more plausible occurrence.

The Gospel of the Holy Twelve claims that one of the primary reasons Jesus was so adamantly condemned by the religious authorities of Israel was because he advocated an end to blood sacrifices at the Temple. To bring an end to these sacrifices, of course, would have completely undermined the financial livelihood of much of the Temple priesthood, and they would have seen Jesus as embodying a personal threat of great consequence. In effect, he disrupted their financial and spiritual foundation; an act more certain to elicit intense opposition from the Judaic priesthood than could scarcely be imagined.

The text also claims, not that Jesus was the "Only Begotten Son", but, phrasing it quite differently, that he was the "*First Begotten Son*" of God. The small change in terminology entirely undermines the traditional church's position that Jesus was a unique Divine being who simply chose to become human; instead, this text now suggests, he was also, in some respects, a human who, through persistent effort and faithfulness to "The Law" (perhaps over many lifetimes), had become a Divine being, suggesting the very gnostic notion that anyone can also attain the same accomplishment.

In a more modern perspective, the text also directly advocates euthanasia, but only in cases of extreme suffering. Always and everywhere throughout the text, the image of Jesus is one utterly dedicated to gentleness and loving care for all beings. Many scenes involved Jesus rebuking someone for cruelly or inflicting pain on others beings, whether people or animals.

Fulfilling "The Law" Within

The Gospel of the Holy Twelve declares that in order to achieve eternal life, "The Law" must be fully obeyed. In this respect this text shows us a very "Essene" Jesus indeed, with his unequivocal focus on "The Law" that must be obeyed. But "The Law", to this Jesus, was not altogether the same Law written in the Hebrew Old Testament, but rather a universal Law pre-written into the inner being of Man. The Law given by Moses, this Jesus claimed, had been altered, betrayed and adulterated by the priests of Persia during the Jewish people's captivity there. The true Law given by Moses was, this scripture maintains, the same ancient Law that is pre-written in the hearts of all men - the "Law of Love and the unity of all life in the One-Family of the All-Parent".

This work teaches that living in accordance to the inner Law is the key to salvation, Eternal

Life, and the Kingdom of Heaven. It teaches that if one experiences the *death of the soul*, it is not because one was condemned by God or anyone else, but by being self-condemned. Whatever the evil doers suffer after death would be that which they themselves created in their own unconscious souls prior to their deaths by betraying the Law, the sense of right and wrong, that is pre-written into our inner being.

Making the Two into One

The gnostic "Gospel of Thomas", "Gospel of Philip", and "Gospel of Truth" found at Nag Hammadi, reconciling and integrating the dual nature of all being is a main focus of this text. God is repeatedly called not "Father", but the Father-Mother, or the All-Parent. His attributes are repeatedly described with equal-but-opposite word pairings such as "Love and Wisdom", "head and heart", "soul and spirit", "within and without", "right and left", and "male and female", or the "Oneness" of the divine pair. But by whatever name, is constantly being mentioned, advocated, and described and declares that salvation comes through the reconciliation and integration of these two primordial elements of being.

The Complete Text

This ancient manuscript claims in no uncertain terms to be the same work composed by the 12 apostles, and, in fact, it makes an intriguing and compelling case for being just that. Its antiquity seems beyond question, as this 19th century text contains words, phrases, and concepts identical to those found in the Dead Sea Scrolls, and the Gnostic Gospels of Nag Hammadi, which were only unearthed in the 1940's. The text therefore cannot, as these connections prove, be anything but authentic.

The Nazarene Way of Essenic Studies contains the full 19th century translation of *The Gospel of the Holy Twelve*, also known as the Gospel of the Nazirenes, the Gospel of the Hebrews, the Original Gospel, the Gospel of the Ebionites, and the Aramaic Matthew. The very Gospel that was repeatedly mentioned and described by many commentators in the early Church as the original teachings of the Nazarene called Christ.

The Nazarene Way of Essenic Studies

Physical Descriptions of Jesus

The Oldest Views and Literary Data on the External Appearance of Jesus the Nazarene

"There is no description of Jesus in the New Testament or in any contemporary source. Yet, in hundreds of icons, paintings, and even coins, there is a common quality that enables us to identify Jesus in works of art. Starting in the sixth century, artistic depictions of Jesus seem inspired or even copied from a single source."

The Oldest Views and Data on the External Appearance of Jesus.
The Apocrypha and Pseudepigrapha (§ 1).
The Church Fathers (§ 2).
Other Data (§ 3). Literary Data on the Oldest Pictures of Jesus.
Extant Pictures of Jesus.
Portraits Ostensibly Authentic.
Portraits by Painters, Sculptors, etc. (§ 1).
Alleged Supernatural Pictures (§ 2). Pictures of Jesus in Ancient Art.
Symbolical and Allegorical Representations (§ 1).
Representations as Teacher and Lawgiver (§ 2). Origin of the Pictures of Jesus.
Iconoclasm: The Religious and Political Destruction of Sacred Images or Monuments.

The Oldest Views and Literary Data on the External Appearance of Jesus

The Description of Publius Lentullus

The following was taken from a manuscript in the possession of Lord Kelly, and in his library, and was copied from an original letter of Publius Lentullus at Rome. It being the usual custom of Roman Governors to advertise the Senate and people of such material things as happened in their provinces in the days of Tiberius Caesar, Publius Lentullus, President of Judea, wrote the following epistle to the Senate concerning the Nazarene called Jesus.

"There appeared in these our days a man, of the Jewish Nation, of great virtue, named *Yeshua* [Jesus], who is yet living among us, and of the Gentiles is accepted for a Prophet of truth, but His own disciples call Him the *Son of God*- He raiseth the dead and cureth all manner of diseases. A man of stature somewhat tall, and comely, with very reverent countenance, such as the beholders may both love and fear, his hair of (the colour of) the chestnut, full ripe, plain to His ears, whence downwards it is more orient and curling and wavering about His shoulders. In the midst of His head is a seam or partition in His hair, after the manner of the Nazarenes. His forehead plain and very delicate; His face without spot or wrinkle, beautified with a lovely red; His nose

The Gospel Of The Holy 12

and mouth so formed as nothing can be reprehended; His beard thickish, in colour like His hair, not very long, but forked; His look innocent and mature; His eyes grey, clear, and quick- In reproving hypocrisy He is terrible; in admonishing, courteous and fair spoken; pleasant in conversation, mixed with gravity. It cannot be remembered that any have seen Him *Laugh,* but many have seen Him *Weep.* In proportion of body, most excellent; His hands and arms delicate to behold. In speaking, very temperate, modest, and wise. A man, for His singular beauty, surpassing the children of men"

The letter from Pontius Pilate to Tiberius Caesar

This is a reprinting of a letter from Pontius Pilate to Tiberius Caesar describing the physical appearance of Jesus. Copies are in the Congressional Library in Washington, D.C.

TO TIBERIUS CAESAR:

A young man appeared in Galilee preaching with humble unction, a new law in the Name of the God that had sent Him. At first I was apprehensive that His design was to stir up the people against the Romans, but my fears were soon dispelled. Jesus of Nazareth spoke rather as a friend of the Romans than of the Jews. One day I observed in the midst of a group of people a young man who was leaning against a tree, calmly addressing the multitude. I was told it was Jesus. This I could easily have suspected so great was the difference between Him and those who were listening to Him. His golden colored hair and beard gave to his appearance a celestial aspect. He appeared to be about 30 years of age. Never have I seen a sweeter or more serene countenance. What a contrast between Him and His bearers with their black beards and tawny complexions! Unwilling to interrupt Him by my presence, I continued my walk but signified to my secretary to join the group and listen. Later, my secretary reported that never had he seen in the works of all the philosophers anything that compared to the teachings of Jesus. He told me that Jesus was neither seditious nor rebellious, so we extended to Him our protection. He was at liberty to act, to speak, to assemble and to address the people. This unlimited freedom provoked the Jews -- not the poor but the rich and powerful.

Later, I wrote to Jesus requesting an interview with Him at the Praetorium. He came. When the Nazarene made His appearance I was having my morning walk and as I faced Him my feet seemed fastened with an iron hand to the marble pavement and I trembled in every limb as a guilty culprit, though he was calm. For some time I stood admiring this extraordinary Man. There was nothing in Him that was repelling, nor in His character, yet I felt awed in His presence. I told Him that there was a magnetic simplicity about Him and His personality that elevated Him far above the philosophers and teachers of His day.

Now, Noble Sovereign, these are the facts concerning Jesus of Nazareth and I have taken the time to write you in detail concerning these matters. I say that such a man who could convert water into wine, change death into life, disease into health; calm the stormy seas, is not guilty of any criminal offense and as others have said, we must agree -- truly this is the Son of God.

Your most obedient servant,
Pontius Pilate

The Emerald of Caesar

This Likeness of Jesus was copied from a portrait carved on an emerald by order of Tiberius Caesar, which emerald the Emperor of the Turks afterwards gave out of the Treasury of Constantinople to Pope Innocent VIII for the redemption of his brother, taken captive by the Christians.

"The Archko Volume"

Another description of Jesus is found in "The Archko Volume" which contains official court documents from the days of Jesus. This information substantiates that He came from racial lines which had blue eyes and golden hair. In a chapter entitled "Gamaliel's Interview" it states concerning Jesus (Yeshua) appearance:

> "I asked him to describe this person to me, so that I might know him if I should meet him. He said: 'If you ever meet him [*Yeshua*] you will know him. While he is nothing but a man, there is something about him that distinguishes him from every other man. He is the picture of his mother, only he has not her smooth, round face. His hair is a little more golden than hers, though it is as much from sunburn as anything else. He is tall, and his shoulders are a little drooped; his visage is thin and of a swarthy complexion, though this is from exposure. His eyes are large and a soft blue, and rather dull and heavy....' This Jew [Nazarite] is convinced that he is the Messiah of the world. ...this was the same person that was born of the virgin in Bethlehem some twenty-six years before..."

- The Archko Volume, translated by Drs. McIntosh and Twyman of the Antiquarian Lodge, Genoa, Italy, from manuscripts in Constantinople and the records of the Senatorial Docket taken from the Vatican of Rome (1896) 92-93

Josephus, the "Antiquities Of The Jews"

This is a quote from Josephus, from his historical first-century writings entitled, "Antiquities Of The Jews," Book #18, Chapter 2, section 3.

> "Now there was about this time Jesus, a wise man, if it be lawful to call him a man; for he was a doer of wonderful works, a teacher of such men as receive the truth with pleasure. He drew over to him both many of the Jews and many of the Gentiles. He was [the] Christ. And when Pilate, at the suggestion of the principal men amongst us, had condemned him to the cross, those that loved him at the first did not forsake him; for he appeared to them alive again the third day, as the divine prophets had foretold these and ten thousand other wonderful things concerning him. And the tribe of Christians, so named from him, are not extinct at this day."

Cornelius Tacitus, a Roman historian

Cornelius Tacitus was a Roman historian who lived circa 56-120 AD. He is believed to have been born in France or Gaul into a provincial aristocratic family. He became a senator, a consul, and eventually governor of Asia. Tacitus wrote at least four historic treatises. Around 115 AD, he published Annals in which he explicitly states that Nero prosecuted the Christians in order to draw attention away from himself for Rome's devastating fire of 64 AD. In that context, he mentions Christus who was put to death by Pontius Pilate.

Christus: Annals 15.44.2-8

"Nero fastened the guilt and inflicted the most exquisite tortures on a class hated for their abominations, called Christians by the populace. Christus, from whom the name had its origin, suffered the extreme penalty during the reign of Tiberius at the hands of one of our procurators, Pontius Pilatus, and a most mischievous superstition, thus checked for the moment, again broke out not only in Judaea, the first source of the evil, but even in Rome..."

Images and Pictures of Jesus

1. The Apocrypha and Pseudepigrapha.

Neither the New Testament nor the writings of the earlier post-Biblical Christian authors have any statements regarding the personal appearance of Jesus, thus contrasting sharply with the Apocrypha and the Pseudepigrapha and especially with the works of the Gnostics. In the "Shepherd" of Hermas (ix. 6, 12) the lofty stature of the Son of God is emphasized, and according to the Gospel of Peter he even towered above the heaven at his resurrection. Gnostic influence is betrayed by visions in which Christ appears as a shepherd, or the master of a ship, or in the form of one of his apostles, as of Paul and of Thomas, or again as a young boy. In the Acts of Andrew and Matthew he assumes the figure of a lad, and the same form is taken in the Acts of Peter and Andrew, in the Acts of Matthew, and in the Ethiopic Acts of James. Manazara is healed by a youth in the Acts of Thomas, and a beautiful lad appears to Peter and Theon in the *Actus Vercellensis*, which also mentions the smile of friendship in the face of Jesus. A handsome youth with smiling face appears at the grave of Drusiana in the Acts of John, but certain widows to whom the Lord restored their sight saw him an, aged man of indescribable appearance, though others perceived in him a youth, and others still a boy. The youthfulness of Christ is also mentioned in the life and passion of St. C☐lus and the vision of Saints Perpetua and Felicitas ascribed to the risen Christ the face of a youth with snow-white hair.

2. The Church Fathers.

The early Christian authors were by no means concordant in their opinions of the personal appearance of Jesus. Some, basing their judgment on Isa. Iii. and liii., denied him all beauty and comeliness, while others, with reference to Ps. xlv. 3, regarded him as the most beautiful of mankind. To the former class belong Justin Martyr, Clement of Alexandria, Basil, Isidor of Peluaium, Theodoret, Cyril of Alexandria, Tertullian, and Cyprian. Origen declared that Christ assumed whatever form was suited to circumstances. It was not until the fourth century that Chrysostom and Jerome laid emphasis upon the beauty of Jesus.

While Isidor of Pelusium had referred the phrase, "Thou art fairer than the children of men" in Ps. xlv. 2, to the divine virtue of Christ, Chrysotom interpreted the lack of comeliness mentioned in Isa. liii. 2 as an allusion to the humiliation of the Lord. Jerome saw in the profound impression produced by the first sight of Jesus upon disciples and foes alike a proof of heavenly beauty in face and eyes. From the insults inflicted upon Jesus Augustine concluded that he had appeared hateful to his persecutors, while actuallly he had been more beautiful than all, since the virgins had loved him.

3. Other Data.

The Problem of the life passion of St. C? us, and the external appearance of Jesus possessed but minor interest for the Church Fathers, although the Catholic Acts of the Holy Apostles ascribe to him an olive complexion, a beautiful beard, and flashing eyes. Further details are first found in a letter to the Emperor Theophilus attributed to John of Damascus (in *MPG*, xcv. 349), which speaks of the brows which grew together, the beautiful eyes, the prominent nose, the curling hair, the look of health, the black beard, the wheat-colored complexion, and the long fingers, a picture which almost coincides with a hand-book on painting from Mt. Athos not earlier than the sixteenth century. In like manner, Nicephorus Callistus, who introduced his description of the picture of Christ (*MPG*, cxlv. 748) with the words, "as we have received it from the ancients," was impressed with the healthful appearance, with the stature, the brown hair which was not very thick but somewhat curling, the black brows which were not fully arched, the sea-blue eyes shading into brown, the beautiful glance, the prominent nose, but brown beard of moderate length, and the long hair which had not been cut since childhood, the neck slightly bent, and the olive and somewhat ruddy complexion of the oval face. A slight divergence from both these accounts is shown by the so-called letter of Lentulus, the ostensible predecessor of Pontius Pilate, who is said to have prepared a report to the Roman Senate concerning Jesus and containing a description of him. According to this document Christ possessed a tall and handsome figure, a countenance which inspired reverence and awakened love and fear together, dark, shining, curling hair, parted in the center in Nazarene fashion and flowing over the shoulders, an open and serene forehead, a face without wrinkle or blemish and rendered more beautiful by its delicate ruddiness, a perfect nose and mouth, a full red beard of the same color as the hair and worn in two points and piercing eyes of a grayish-blue.

II. Literary Data on the Oldest Pictures of Jesus:

(1) A handkerchief embroidered with the figures of Jesus and his Apostles, and made, according to legend, by his mother, is said to have been seen by the monk Arculfus during his residence in Jerusalem (Adamnan, *De Locis sanctis*, i. 11 [12]). (2) In his account of his visit to C☐rea Philippi, Eusebius mentions (*Hist. eccl.* vii. 18) a group of statuary in brass which consisted of a kneeling woman and a man standing with his hands stretched out toward her. Local tradition saw in this a figure of Jesus and the woman healed of an issue of blood, who was said to have come from C☐rea Philippi. This legend was accepted by Eusebius, Asterius Amasenus Photius, Sozomen, Philostorgius, and Macarius Magnes, the last-named calling the woman Beronike. The actual meaning of the group is uncertain. Some have seen in it an emperor and a province, possibly Hadrian and Judea while others have regarded it as ☐culapius and Hygeia, a view which is vitiated by the fact that no mention is made of the serpent-staff characteristic of statues of the god of healing. It is entirely possible that the group actually represented Christ and either the woman with an issue of blood or possibly the woman of Canaan who implored him to heal her daughter. (3) According to Iren☐ (*H☐/i>.*, I., xxv. 6), pictures of Christ were

possessed by the Gnostic sect of Carpocratians, who crowned them with garlands like the pictures of philosophers--Pythagoras, Plato, Aristotle, and others--while, according to the Carpocratians, Pilate had a portrait of Jesus painted during his lifetime, and the Carpocratian Marcellina possessed a picture of Christ which she honored, like those of Paul, Homer, and Pythagoras, with prayer and incense. (4) The Emperor Alexander Severus had a picture of Jesus; it must have been, however, only an ideal portrait, like those of Apollonius, Abraham, Orpheus, and others, which were also included in his lararium (Lampridius, Vita Alex. Sev. xxix.). (5) A brass statue of the Savior was erected by Constantine the Great before the main door of the imperial palace of Chalce (Theophanes in MPG, cviii. 817). (6) A picture of Jesus "painted from life" was possessed by the Archduchess Margaret which may be the same one as D□□'s altar-piece of St. Luke at Brussels (M. Thausing, D□□, p. 420, Leipsic, 1876).

While the portraits just mentioned were prepared by human agency, there were others to which a supernatural origin was ascribed. To this category belong (7) a picture at Camulium in Cappadocia, apparently on cloth and perhaps a copy of that of Edessa (see below). It was mentioned at the second Nicene Council and was carried to Constantinople by Justin II., where it was regarded as so sacred that a special festival was instituted in its honor, and it was frequently carried in war as a potent icon (J. Gretsei opera, xv. 196-197, Regensburg, 1741). (8) In the war against the Persians the General Philippicus had a picture of Christ which the Romans believed to be supernatural in origin, and the same portrait served to quell a mutiny in the army of Priscus, the successor of Philippicus. This icon was apparently on cloth, and was a copy of an original which was frequently confounded with a portrait in Amida, although the latter is expressly said to have been painted, and was, consequently, natural in provenience (Zacharias, MPG, lxxxv. 1159). (9) A Syriac fragment mentions a picture of Jesus painted on linen and found unwet in a spring by a certain Hypatia shortly after the Passion. This portrait left a miraculous imprint on the napkin in which it was wrapped, and one of these pictures found its way to C□rea while the other was taken to Comolia (possibly identical with the city of Camulium already mentioned), although a copy was later found at Dibudin (?) (Lipsius, Die edessenische Abgarsage, p. 67, n. 1, Brunswick, 1880). (10) About 570 a linen mantle was shown at a church in Memphis which bore the impress of the Savior's face and was so bright that none could gaze at it (Antoninus Martyr, De locis sanctisxliv.). (11)Byzantine literature frequently mentions pictures of Christ impressed on bricks. According to a legend which presents several slight variations, the portrait of himself which Jesus had sent to Abgar at Edessa was believed to have been walled up to save it from the attack of King Ananun and to have been rediscovered in 539 together with a brick which bore a miraculous copy of the original (Georgius Cedrenus, ed. Bekker, i. 312, and others). (12) The patriarch Germanus, when forced to leave Constantinople, is said to have taken with him a picture of Christ which later came into the possession of Gregory II. (G. Marangoni, Istoria dell' oratorio di San Lorenzo, pp. 78 sqq., Rome, 1747). (13) The cloth with a picture of Christ presented by Photius to the hermit Paul at Latro in the ninth century was merely a copy of a miraculous original, although only he to whom the gift was made was able to perceive the portrait, others seeing only the cloth (Gretses, ut sup. p,186). (14) More important than all other statements concerning the oldest pictures of Christ is a passage of Augustine (De trin. viii. 4), stating that the portraits of Jesus were innumerable in concept and design.

III. Extant Pictures of Jesus.

1. Portraits Ostensibly Authentic:
1. Portraits by Painters, Sculptors, etc.

(1) The paintings of Luke, of which the best known are two at Rome. One of these is in the chapel Sanctus Sanctorum, although the statement that Luke painted a portrait of Jesus dates only from medieval times, the monk Michael, the biographer of Theodore of Studium, being one of the earliest sources. In the last quarter of the twelfth century the legend of Luke was interwoven by Wernher of Niederrhein with the tradition of Veronica (see below). Luke, in answer to Veronica's entreaties, is said to have made repeated attempts to portray Christ, but his endeavors were unsuccessful. Jesus then impressed the image of his face upon the handkerchief of Veronica. Another picture ascribed to Luke and painted on cloth is in the Vatican library, while a third is said to have been placed in the cathedral of Tivoli by Pope Simplicius. Other pictures are likewise ascribed to a similar provenience, and very late traditions even attribute statues of Christ to the chisel of Luke. [In the church of San Miniato at Monto, in the environs of Florence, Italy, is shown a portrait of Christ, attributed to Luke.] (2) To Nicodemus is ascribed a statue of the crucified Christ carved in black cedar and preserved in the Cathedral of Lucca. Its design shows that it dates at the earliest from the eighth century, although tradition states that the model of Nicodemus was furnished by the impress of the Savior's body on the linen cloths purchased to cover the corpse at the descent from the cross. (3) A "true and only portrait of our Savior taken from an engraved emerald which Pope Innocent VIII. received from Sultan Bajazted II. for the ransom of his brother, who was a captive of the Christians," frequently reproduced in photograph is in reality the copy of a medal which may have been cut at the command of Mohammed II., and which is, at all events, of comparatively modern date. (4) The mosaic in the Church of St. Praxedis in Rome, which is exhibited on festal occasions, is by no means one of the earliest Christian mosaics, although tradition regards it as a present to Pudens from the Apostle Peter.

2. Alleged Supernatural Pictures.

Alleged supernatural pictures may be divided into those which represent the entire figure of Jesus, and those which give only his face. (1) Clothe of medieval date containing more or less clear outlines of the figure of a man, all claiming to be the "napkin" in which Jesus was wrapped in the grave and on which his image was impressed, were formerly found in Chamb□, and until the end of the eighteenth century, in Besan□, while they still exist at Compi□e and Turin, the latter "napkin" being declared authentic by a bull of Sixtus IV. Far more famous, however, are the cloths which bear only the impress of a head or face and of these one of the best known is (2) the picture of Edessa, or the Abgar picture. According to the Doctrine of Addai and Moses of Choren, Hanan, the envoy of the king of Edessa, painted a portrait of Jesus and took it to his royal master. Evagrius, on the authority of Procopius, states that Christ sent to the king a picture of miraculous origin. The legend apparently arose about 350, and may well have been based on an actual painting which remained at Edessa till 944, when it was brought to Constantinople by the Emperor Romanus I. Its subsequent fortunes are uncertain, although various cities laid claim to its possession, especially Genoa, Rome, and Paris, the first-named city advancing the most probable arguments for authenticity and receiving the confirmation of Pius IX. (see ABGAR). This picture shows only the head of Jesus, but legend also knows a full-length Edessene portrait on linen produced by contact with the body of Christ. It is mentioned by Gervase of Tilbury in the beginning of the thirteenth century, who bases his statement on ancient sources and says that it was exhibited on festivals in the chief church of Edessa, and that on Easter it shows Jesus successively as a child, boy, youth, young man, and in the ripeness of years. (3) One of the choicest treasures of the Roman Church is the handkerchief of Veronica, which is shown only on special occasions, particularly in Passion Week. This portrait is said to have been transferred in 1297 by Boniface VIII. from the Hospital of the Holy Ghost to St. Peter's in Rome, where it reposes behind the statue of St. Veronica. The picture, which is now much faded, shows an elliptical face with a low-arched forehead, in marked contrast with the long nose. The mouth is slightly open, and the scanty hair is visible only on the temples. The beard on the cheeks is thin, but is stronger on the chin, where it ends in three points, while the mustache is more conspicuous for color than for strength. The eyes arched by scanty brows, are closed, and, combined with features distorted by agony and stained with blood complete the picture of a martyr pale in death. From the point of view of esthetics and the history of art, the picture is probably Byzantine. Although one would expect the picture of Veronica to be regarded as the napkin which covered the head of Christ, there is no tradition as to its origin, although a mess of medieval legends connects it with the name of a woman.

These may be divided into two classes. In the older group, apparently written shortly before the ninth century, Veronica appears as the woman afflicted with an issue of blood, who had a portrait of Jesus either painted by herself or at her bidding, or else impressed by Christ himself upon a piece of cloth. The second form of the legend sprang up in France and Germany in the course of the fourteenth century and superseded the older version before 1500. According to this tradition, Veronica gave the Savior a handkerchief on his way to Golgotha, and received it back impressed with his features. Further amplifications of the tradition stated that the napkin was brought to Rome by John VII., or even during the reign of Tiberius, while it is certain that Celestine III. prepared a reliquary for it. At all events, what is clear is that during the medieval period Rome possessed a cloth picture of Christ, which was apparently supposed to be the miraculous impress of the head of Jesus in the sepulcher. It is significant, moreover, that it bore the name sudarium before the rise of the legend of the handkerchief given Christ to wipe his face on his way to the cross, nor was it until the twelfth century that the name of Veronica even began to form a part of the tradition, a connection suggested by a popular etymology of Veronica as Vera *?* ("true image"), This legend of Veronica gave rise to a tendency of art which reached its culmination in D☐☐, who represented the napkin of Veronica and the Savior with a crown of thorns, combining the suffering in the face of Jesus with the loftiness and the majesty of the Son of God, (4) The picture of Christ in the apse of St. John Lateran at Rome is supposed to have been miraculously produced when the church was dedicated by Pope Sylvester, although it is in reality a mosaic of recent date.

2. Pictures of Jesus in Ancient Art:
1. Symbolical and Allegorical Representations.

In the course of time pictorial representations of Jesus became either real or symbolical and allegorical, the latter tendency gradually giving way to the former. To the category of symbols belong the fish, the lamb, the various monograms of Christ, and the Good Shepherd, the last-named leading to representations of Jesus in human form. As early as Tertullian the Good Shepherd adorned chalices, and it was a favorite form of decoration in the catacombs, where the figure usually carries a goat or a wether. In these pictures, often adorned with other animals, trees, and shrubs, and based on Luke xv. 5; John x.; and Ps. xxiii., the Christ appears only in youthful guise, although the Shepherd is usually clad in garments of a higher rank and wears the Roman tunic and the pallium as well as sandals. The figure, moreover, is Latin instead of Oriental in type, and represents a youthful and beardless sometimes even boyish, figure, a round head with curling hair, and a frank face with regular features. This type of picture, purely ideal as it was, underwent evolution in the course of time. In the third century the face grew more oval, while the unparted hair grew slightly over the forehead in the center and flowed on the on the sides in wavy or curly locks.

2. Representation as Teacher and Lawgiver.

The first real impulse, however, to artistic representations of Jesus was given by his miracles, though the risen Lord as a teacher and a lawgiver became more and more a subject for pictorial representation. In the midst of all or a part of his disciples, including Paul, Christ appears either on a plain, as in Spain and southern France, or standing on a mountain either within or without the four rivers of Eden, or sitting on a throne with his feet on a footstool or on the clouds while mosaics represent him as seated on the celestial globe. As a teacher, he is depicted as speaking and as holding a book or scroll either in his hand or on his bosom, while as a lawgiver he proffers the Gospel to Peter or Paul. In both of these latter categories the beardless, youthful type gradually grows less frequent, so that on Roman, Upper Italian, and French sarcophagi the central Christ appears bearded, although in the reliefs on their sides he wears no beard, the former representing the risen Lord and the latter the earthly Savior. Originally a characteristic of the ascended Christ, the beard was attributed to Jesus during his earthly ministry after the end of the fourth or the beginning of the fifth century. The struggle between the two types is seen in the mosaics of Sant'Apollinare Nuovo at Ravenna and of St. Michael, but the earliest specimen of the bearded Christ is generally believed to be the socalled Callistinian mosaic which was found in the catacomb of St. Domitilla. In conformity with the manhood implied by the beard, the body increased in height and breadth, while the features became more sharply defined as the bones gained in accentuation over the flesh. The nose became longer and more prominent, and the eyes were deeper and their pupils enlarged, while the angles of the nose and mouth were more sharply outlined. The hair, while frequently less curling than hitherto, was now represented as falling to the neck and shoulders, and was often parted in the middle. The color both of the hair and of the beard varied through all shades from yellow to gray and black. The upper lip was never clean-shaven, and the beard was sometimes close and sometimes either pointed or rounded, the parted type being found only in rudimentary form in early Christian art.

The bearded Christ represents the climax of the art of early Christianity, and the fifth century ushered in a period of decay marked by all manner of exaggeration. Majesty became stiffness, exaltation unapproachability, and earnestness gloom. Thus the Christ of Saints Cosmas and Damian (q.v.) in Rome, dating from the sixth century, is a figure with, long face, projecting cheek bones, ashen complexion, attenuated nose, mane-like hair, and scanty beard.

It was the task of the Middle Ages to reduce the multiplicity of concepts of the likeness of Christ to unity, a task which required centuries for its completion. The Carolingian period saw a sort of fruitless recrudescence of the process of evolution of the early Christian Period. Even during the Renaissance the beardless type struggled for supremacy with the bearded, especially in miniatures and ivories, but the former steadily lost ground, so that its last sporadic occurrence is a Scandinavian Christ in glory of the thirteenth century, such pictures as the Piet□i> of Botticelli at Munich being mere anachronisms.

IV. Origin of the Pictures of Jesus:

While the theory may be advanced that the oldest pictures of Christ were based either on works of art still more ancient or on tradition, it is practically certain that they are not real portraits but ideal representations. This is clear both from their extreme diversity and from the words of Augustine: "What his appearance was we know not." The most primitive type, wherein early Christian and Gnostic documents agree, is that of a boy or youth. The youthful vigor of the early Church in religious and in moral thought, sustained by the belief in the second coming of the Lord and strengthened by persecution, inspired the artist to depict the Christ as the incarnation of undying youth, even as Noah, Job, Abraham, and Moses were represented as beardless boys. Herein, too, lay the genesis of the concept of the Good Shepherd.

With the fourth and fifth centuries the bearded type was evolved side by side with the beardless. The explanation of this change lies in the perfection, strength, and manliness implied by the beard. The parted hair, on the other hand, which is characteristic of the pictures of Christ in this period, especially in the mosaics, typifies his earthly lineage and designates him as one of the children of Israel, since of human beings only Jews and Judeo-Christians are represented with parted hair in early Christian art. The theory, advanced by many scholars, that Greek religious art influenced the various early Christian concepts of the personal appearance of Christ seems to lack sufficient evidence to be in any wise conclusive.

In the 1930's, French Shroud scholar Paul Vignon described a series of common characteristics visible in many early artistic depictions of Jesus. The Vignon marking, as they are known, all appear on the Shroud suggesting that it is the source of later pictures of Jesus.

Christ Pantocrator, c. 1100 from dome of Church at Daphni, near Athens. Note U at bridge of nose, triangle on nose, raised right eyebrow, uneven hair, owlish eyes.

- A square U-shape between the eyebrows.
- A downward pointing triangle or V-shape just below the U-shape, on the bridge of the nose.
- Two wisps of hair going downward and then to the right.
- A raised right eyebrow.
- Large, seemingly "owlish" eyes.
- An accent on the left cheek and an accent on the right cheek that is somewhat lower.
- A forked beard and hair parted in the middle, a custom of the Nazarenes.
- Hair on one side of the head that is shorter than on the other side.
- An enlarged left nostril.
- An accent line below the nose and a dark line just below the lower lip.
- A gap in the beard below the lower lip.
- Draped clothing of white linen typical of the ancient Essenes.

Iconoclasm
The religious and political destruction of sacred images or monuments

Literally, iconoclasm is religious and political destruction of the sacred images or monuments, usually (though not always) of another religious group. People who destroy such images are called iconoclasts, while people who revere or venerate such images are called iconodules.

In 725 the Emperor Leo III, ignoring the opposition of both Patriarch Germanus of Constantinople and Pope Gregory II in Rome, ordered the removal of all icons from the churches and their destruction. Nearly all ancient images of Jesus were destroyed during the iconoclastic periods in the eighth and ninth centuries.

Table of Contents

1 Byzantine iconoclasm
1.1 The first iconoclastic period: 730-787
1.2 The second iconoclastic period: 813-843
2 Islamic iconoclasm
3 Reformation iconoclasm

Just as in our own time there is controversy about icons, so was there dispute in the early Church. Early critics of icons included Tertullian, Clement of Alexandria, Minucius Felix and Lactancius. Eusebius was not alone in fearing that the art of the pagan world carried with it the spirit of the pagan world while others objected on the basis of Old Testament restrictions of imagery. Christianity was, after all, born in a world in which many artists were employed doing religious or secular work. Idolatry was a normal part of pagan religious life. Thus we find that in the early centuries, in the many areas of controversy among Christians, there was division on questions of religious art and its place in spiritual life.

Byzantine Iconoclasm

The first iconoclastic period: 730-787 Emperor Leo III the Isaurian (reigned 717-741) banned the use of icons of Jesus, Mary, and the Saints and commanded the destruction of these images in 730. The Iconoclastic Controversy was fueled by the refusal of many Christians resident outside the Byzantine Empire, including many Christians living in the Islamic Caliphate to accept the emperor's theological arguments. St. John of Damascus was one of the most prominent of these. Ironically, Christians living under Muslim rule at this time had more freedom to write in defense of icons than did those living in the Byzantine Empire. Leo was able to promulgate his policy because of his personal popularity and military success - he was credited with saving Constantinople from an Arab siege in 717-718 and then sustaining the Empire through annual warfare.

The first Iconoclastic period came to an end at the Second Council of Nicaea in 787, when the veneration of icons was affirmed, although the worship of icons was expressly forbidden. Among the reasons were the doctrine of the Incarnation: because God the Son (Jesus Christ) took on flesh, having a physical appearance, it is now possible to use physical matter to depict God the Son, and to depict the saints. Icon veneration lasted through the reign of Empress Irene's successor, Nicephorus I (reigned 802-811), and the two brief reigns after his.

The second Iconoclastic period: 813-843

Emperor Leo V (reigned 813-820) instituted a second period of Iconoclasm in 813, which seems to have been less rigorously enforced, since there were fewer martyrdoms and public destructions of icons. Leo was succeeded by Michael II, who was succeeded by his son, Theophilus II. Theophilus died leaving his wife Theodora regent for his minor heir, Michael III. Like Irene 50 years before her, Theodora mobilized the iconodules and proclaimed the restoration of icons in 843. Since that time the first Sunday of Lent is celebrated in the churches of the Orthodox tradition as the feast of the "Triumph of Orthodoxy".

Islamic Iconoclasm

Because of the prohibition against figural decoration in mosques - not, as is often said, a total ban on the use of images - Muslims have on occasion committed acts of iconoclasm against the devotional images of other religions. An example of this is the 2001 destruction of frescoes and the monumental statues of the Buddha at Bamiyan by the Taliban, an element of the Islamist movement.

In a number of countries, conquering Muslim armies tore down local temples and houses of worship, and built mosques on their sites. The Dome of the Rock in Jerusalem was built on top of the remains of the Jewish Temple in Jerusalem. Similar acts occurred in parts of north Africa under Muslim conquest. In India, numerous former Buddhist monasteries and Hindu temples were conquered and rebuilt as mosques. In recent years, some Hindu nationalists have attempted to tear down these mosques, and replace them with Hindu Temples. This is part of the current conflict today between Indian Hindu nationalists and Indian Islamists.

Reformation Iconoclasm

Some of the Protestant reformers encouraged their followers to destroy Catholic art works by insisting that they were idols. Huldreich Zwingli and John Calvin promoted this approach to the adaptation of earlier buildings for Protestant worship. In 1562, some Calvinists destroyed the tomb of St. Irenaeus and the relics inside, which had been under the altar of a church since his martyrdom in 202.

The Netherlands (including Belgium) were hit by a large wave of Protestant iconoclasm in 1566. This is called the Beeldenstorm.

Bishop Joseph Hall of Norwich described the events of 1643 when troops and citizens, encouraged by a Parliamentary ordinance against superstition and idolatry, behaved thus:

> *'Lord what work was here! What clattering of glasses! What beating down of walls! What tearing up of monuments! What pulling down of seats! What wresting out of irons and brass from the windows! What defacing of arms! What demolishing of curious stonework! what tooting and piping upon organ pipes! And what a hideous triumph in the market-place before all the country, when all the mangled organ pipes, vestments, both copes and surplices, together with the leaden cross which had newly been sawn down from the Green-yard pulpit and the service-books and singing books that could be carried to the fire in the public market-place were heaped together'.*

BIBLIOGRAPHY: The Nazarene Way of Essenic Studies. A Painter's Study of the Likeness of Christ from the Time of the Apostles, London, 1903; A. N. Didron, Iconographie chr□enne. Histoire de Dieu, Paris, 1843; W. Grimm, Die Sage vom Ursprung der Christusbilder, pp. 121-175, Berlin, 1844; Mrs. Jameson, History of our Lord as Exemplified in Works of Art, 2 vols., London, 1872; A Hauck, Die Entstehung des Christustypus in der abendl□ischen Kunst, Heidelberg, 1880; T. Heaphy, Likeness of Christ, New York, 1886 (illustrations valuable); H. M. A. Guerber, Legends of the Virgin and Christ, with Special Reference to . . . Art, ib. 1896; E. M. Hurll, Life of Our Lord in Art, Boson, 1898 (valuable); E. von Dobsch□□ Christusbilder, Leipsic, 1899; F. W. Farrar, Life of Christ as Represented in Art, London, 1900; J. L. French, Christ in Art, Boston, 1900; F. Johnson, Have We the Likeness of Christ, Chicago, 1903: J. Burns, The Christ Face in Art, New York, 1907; J. S. Weis-Liebersdorf, Christus- und Apostelbilder, Freiburg, 1902; J. Heil, Die fr□□ristlichen Darstellungen der Kreuzigung Christi, Leipsice, 1904; K. M. Kaufmann, Handbuch der christlichen Arch□ogie, Paderborn, 1905; G. A. M□□r, Die liebliche Gestalt Jesu Christ, nach der schriftlichen und monumentalen Urtradition, Styria, 1909.

Physical Description of Jesus

Christ is shown using compasses to re-enact the creation of the universe from the chaos of the primal state. This icon can also be understood as an image of individual self-creation; for here, as in many medieval images of Christ, Tantric symbolism is evident. Christ holds the compass with his hand across the vital centre called the heart chakra, and from this centre he organizes the turmoil of the vital energies contained in the lower chakras which are indicated on the body by centres at the navel and genitals. Geometry is symbolized here in both the individual and universal sense as an instrument through which the higher archetypal realm transmits order and harmony to the vital and energetic worlds.

(words on right hand side by Robert Lawlor, from his classic book " Sacred Geometry").

Testimony of the Early Christian Church
to the Universal Abstinence From Flesh Diet and Strong Drink

WHEN St. John Chrysostom (D. *407),* in his homily on Matthew xxii, 1-14, tells us that "flesh-meats and wine serve as materials for sensuality, and are a source of danger, sorrow, and disease," he does not stand alone.

Writing, in confutation of Jovinian, a monk of Milan, who abandoned asceticism, St. Jerome (D. *A.V.* 440) holds up vegetarianism as the Christian ideal and the restoration of the primeval rule of life. The passage may be rendered :--" As to his argument that in God's Second Blessing permission was given to eat flesh- a permission not given in the first Blessing- let him know that just as permission to put away a wife was, according to the words of the Saviour, not given from the beginning, but was granted to the human race by Moses because of the hardness of our hearts. So also in like manner the eating of flesh was unknown until the flood, but after the Flood, just as quails were given to the people when they murmured in the desert, so have sinews and the offensiveness of flesh been given to our teeth. The Apostle, writing to the Ephesians, teaches us that God had purposed that in the fullness of time he would restore all things, and would draw to their beginning, even to Christ Jesus, all things that are in heaven or that are on earth. Whence also, the Saviour Himself, in the Apocalypse of John, says, ' I am Alpha and Omega, the beginning and the end.' From the beginning of human nature, we neither fed upon flesh nor did we put away our wives, nor were our foreskins taken away from us for a sign. We kept on in this course until we arrived at the Flood. But after the Flood, together with the giving of the Law, which no man could fulfill, the eating of flesh was brought in; and the putting away of wives was conceded to hardness of heart; and the knife of circumcision is brought into use; as if the hand of God had created in us more than is necessary. But now that Christ has come in the end of time, and has turned, back Omega to Alpha, and drawn back the end to the beginning, neither is it permitted to us to put away our wives, nor are we circumcised, nor do we eat flesh; hence the Apostolic saying, ' It is a good, thing not to drink wine, and not to eat flesh.' For wine also, together with flesh, began to be used after the Flood."

Not less striking is the testimony of St. Basil (D. 379) : " With sober living," he says, " well-being increases in the household, animals are in safety, there is no shedding of blood, nor putting animals to death. The knife of the cook is needless; for the table is spread only with the fruits that Nature gives, and with them they are content. John the Baptist, he continues, "had neither bed, nor table, nor inheritance, nor ox, nor grain, nor baker, nor other things regarded as the necessaries of life; and yet it was to him that the Son of God gave the eulogy that he was the greatest of the children of men."

The Gospel according to the Hebrews was that which was in use amongst the first Christians of Jerusalem, and the Gospel according to the Egyptians is thought to have been in close relation to it. It has been said that there are traces of it in the Talmud before A.D. 130., It has even been conjectured that it was the Hebrew source from which the present Gospel according to Matthew was derived. This Gospel, according to the Nazarenes, was widely. circulated in the early Church, and was held in high esteem by the Jewish Christians.

Hegesippus gives a remarkable account of James, the brother of the Lord, and the first ruler of the Christian Church in Jerusalem. James, we are told was Holy from birth. He drank no wine nor strong liquor, nor ate he any living thing. A razor never went upon his head, and neither used the bath nor anointing with oil. Even his clothes were free from any taint of death for he wore no woolen but linen garments only., " It is a remarkable fact that Instead of being represented as a sectary at the head of a new school of religious thought antagonistic to the ancient Hebrew faith, we are told that he, and he alone, was permitted to enter the sanctuary.

That the physical puritanism of abstainence from intoxicants and flesh-meats was not an ideal foreign to Judaism we know from the examples of the Rechabites, the Nazarites, the Nazarenes, and the Essenes. The accounts that have come down to us of the last named sect are very interesting and suggestive. They lived in a brotherly community, they cultivated the land, they observed the Sabbath strictly, they refused to swear, they abstained from intoxicants and flesh.

There are striking parallelisms between Essenism and Christianity. Seek first the kingdom of God was the aim of the Essenes (Matt, vi, 33, Luke, xii, 31). Sell your possessions and give to the poor (Matt vi, 33)., They despised riches (Matt vi, 19-21). The brotherly spirit amongst them was a wonder to the Jewish people, and a test of Christianity is "we know that we have passed from death to life because we love the brethren" (I John iii, 14). The Essenes and the Christians in Jerusalem lived in communities where each man had a share in the common. No wonder that De Quincey with his love of paradox should declare the Essenes to be "neither more or less than the new-born brotherhood of Christians.

The writer of these few extracts makes acknowledgment the same to E. A.' Axon, LL.D., F.R.S.L.

The Nazarene Way of Essenic Studies
The Epistle of Apollos The Prophet

APOLLOS TO HIERASTHENES, GREETING.-

1 TOUCHING the matter whereby thou didst enquire in thy last epistle, I will inform thee even as I have received. I Appollos was in my house in Nazareth after the Holy City had been taken by the Romans, and the Temple of God destroyed, even as the Lord had told us.

2 And as the sun went down and I was resting from my work, the room was filled with a bright light and there appeared unto me Agella., my sister (who had been reported as dead with many others of the brethren who were in the Holy City at the time of the siege and who have never since been seen by any to this day).

3 And Agella spake to me saying, BROTHER, why grievest thou for me, and for the fall of Ierusalem and for the Holy House. Grieve rather that thou wast left behind when we with others of the brethren who were ready were taken up from the earth.

4 For when the city was sorely besieged and the battle was the most fierce and the confusion great and terrible, there was seen by all a great wonder in the heavens.

5 For the Lord himself appeared from the clouds **with her to whom he first appeared after he rose from the dead**, who announced his resurrection to the twelve, and the holy angels, according to the word that he Spake unto us while he was in the flesh.

The Gospel Of The Holy 12

6 And we who sore longed for deliverance and were ready for his appearance were caught up to him in the clouds with Iohn, who alone of the twelve remained (whether in the body or out of the body I knew not).

7 It was in a moment and we were changed in the twinkling of an eye, and those who were his enemies saw it and fled in great confusion and fell on the swords of the Romans and perished, and to me alone has it been given that I should appear unto thee for thy comfort my brother, and for the consolation of those that are left behind and those that shall come after them, **that they may believe in the words spoken by the Lord before he suffered.**

8 Farewell brother, and go and comfort those that are left, for there will arise those who will deny that he returned as he said, because none of those who saw his appearance are left behind to witness thereof.

9 But believe thou that the Christ shall return again at the end of the Age in glory.

10 AND I arose and went to some of the brethren and told them these things, but they seemed to them as an idle tale, for they answered, If thy sister and the others were taken, why have we been left behind in the misery of this world? Surely they have fallen by the sword also, and it was a vision, and we which are left behind shall perish likewise?

11 And I returned to my home and held my peace, for I was in doubt, and said, If the thing is true it will be brought to light in a future day, for the Lord certainly did say that "before this generation should pass away all these things should be," even as my sister hath told me they have been.

12 They who are with me salute thee. Peace be with thee, and to all in thine house.

EPISTLE OF APOLLOS TO HIERASTHENES.

The 96 LECTIONS

The Gospel of the Holy Twelve
Translated from the original Aramaic
by Rev. G.J.R. Ouseley

**IN THE NAME OF
THE ALL HOLY. AMUN.**

Here beginneth the Gospel of the Perfect Life of Iesu-Maria, the Christ, the offspring of David through Ioseph and Mary after the flesh, and the Son of God, through Divine Love and Wisdom, after the Spirit.

Lection 1

The Parentage And Conception Of Iohn The Baptist

1. THERE was in the days of Herod, the King of Judea, a certain priest named Zacharias, of the course of Abia; and his wife was of the daughters of Aaron, and her name was Elisabeth.

2. And they were both righteous before God, walking in all the commandments and ordinances of the Lord blameless. And they had no child, because that Elisabeth was barren, and they both were now well stricken in years.

3. And it came to pass, that while he executed the priest's office before God in the order of his course, according to the custom of the priest's office, his lot was to burn incense when he went into the temple IOVA. And the whole multitude of the people were praying without at the time of the offering of incense.

4. And there appeared unto him an angel of the Lord standing over the altar of incense. And when Zacharias saw, he was troubled, and fear fell upon him. But the angel said unto him, Fear not, Zacharias, for thy prayer is heard; and thy wife Elisabeth, shall bear thee a son, and thou shalt call his name Iohn.

5. And thou shalt have joy and gladness; and many shall rejoice at his birth; for he shall be great in the sight of the Lord, and shall neither eat flesh meats, nor drink strong drink; and he shall be filled with the Holy Spirit, even from his mother's womb.

6. And many of the children of Israel shall he turn to the Lord their God; And he shall go before him in the spirit and power of Elias, to turn the hearts of the fathers to the children, and the disobedient to the wisdom of the just; to make ready a people prepared for the Lord.

7. And Zacharias said unto the angel, Whereby shall I know this? for I am an old man, and my wife is well stricken in years. And the angel answering said unto him, I am Gabriel, that stand in the presence of God; and am sent to speak unto thee, and to announce unto thee these glad tidings.

8. And, behold, thou art dumb, and not able to speak, until the day that these things shall be performed, then shall thy tongue be loosed that thou mayest believe my words which shall be fulfilled in their season.

9. And the people waited for Zacharias, and marvelled that he tarried so long in the temple. And when he came out, he could not speak unto them; and they perceived that he had seen a vision in the temple; for he made signs unto them, and remained speechless.

10. And it came to pass, that, as soon as the days of his ministration were accomplished, he departed to his own house. And after those days, his wife Elisabeth, conceived, and hid herself five months saying, Thus hath the Lord dealt with me in the days wherein he looked on me, to take away my reproach among men.

Lection 1.1 The opening paragraph of this Gospel was evidently before the eyes, or in the mind of St. Paul when he wrote Romans 1-4. (See Luke 1:5) This is only one of several instances where this Gospel, or the words of Iesus recorded in it, are used subsequently, without specially indicating the fact (as shewn further on), being well-known to his hearers at that time.

Lection 2

The Immaculate Conception
Of Iesus The Christ

1. AND in the sixth month the angel Gabriel was sent from God, unto a city of Galilee, named Nazareth, to a virgin espoused to a man whose name was Ioseph, of the house of David; and the virgin's name was Mary.

2. Now Ioseph was a just and rational Mind, and he was skilled in all manner of work in wood and in stone. And Mary was a tender and discerning Soul, and she wrought veils for the temple. And they were both pure before God; and of them both was Iesu-Maria who is called the Christ.

3. And the angel came in unto her and said, Hail, Mary, thou that art highly favoured, for the Mother of God is with thee: blessed art thou among women and blessed be the fruit of thy womb.

4. And when she saw him, she was troubled at his saying, and cast in her mind what manner of salutation this should be. And the angel said unto her, Fear not, Mary, for thou hast found favour with God and, behold, thou shalt conceive in thy womb and bring forth a child, and He shall be great and shalt be called a Son of the Highest.

5. And the Lord God shall give unto him the throne of his father David: and he shall reign over the house of Jacob forever; and of his kingdom there shall be no end.

6. Then said Mary unto the angel, How shall this be, seeing I know not a man? And the angel answered and said unto her The Holy Spirit shall come upon Ioseph thy Spouse, and the power of the Highest shall overshadow thee, O Mary, therefore also that holy thing which shall be born of thee shall be called the Christ, the Child of God, and his Name on earth shalt be called Iesu-Maria, for he shall save the people from their sins, whosoever shall repent and obey his Law.

7. Therefore ye shall eat no flesh, nor drink strong drink, for the child shall be consecrated unto God from its mother's womb, and neither flesh nor strong drink shall he take, nor shall razor touch his head.

8. And, behold, thy cousin Elisabeth, she hath also conceived a son in her old age: and this is the sixth month with her, who was called barren. For with God no thing shall be impossible. And Mary said, Behold the handmaid of the Lord; be it unto me according to thy word. And the angel departed from her.

9. And in the same day the angel Gabriel appeared unto Ioseph in a dream and said unto him, Hail, Ioseph, thou that art highly favoured, for the Fatherhood of God is with thee. Blessed art thou among men and blessed be the fruit of thy loins.

10. And as Ioseph thought upon these words he was troubled, and the angel of the Lord said unto him, Fear not, Ioseph, thou Son of David, for thou hast found favour with God, and behold thou shalt beget a child, and thou shalt call his name Iesu-Maria for he shall save his people from their sins.

11. Now all this was done that it might be fulfilled which was written in the prophets saying, Behold a Maiden shall conceive and be with child and shall bring forth a son, and shall call his name Emmanuel, which being interpreted is, God Within Us.

12. Then Ioseph being raised from sleep did as the angel had bidden him, and went in unto Mary, his espoused bride, and she conceived in her womb the Holy One.

13. AND Mary arose in those days and went into the hill country with haste, into a city of Judea and entered into the house of Zacharias, and saluted Elisabeth.

14. And it came to pass, that, when Elisabeth heard the salutation of Mary, the babe leaped in her womb; and Elisabeth was filled with the power of the Spirit and spake, with a clear voice and said, Blessed art thou among women and blessed is the fruit of thy womb.

15. Whence is this to me, that the mother of my Lord should come to me? For, lo, as soon as the voice of thy salutation sounded in my ears, the babe leaped for joy. And blessed is she that believed: for there shall. be a performance of those things which were told her from the Holy One.

16. And Mary said: My soul doth magnify Thee, the Eternal, and my spirit doth rejoice in God my Saviour. For thou hast regarded the low estate of thy handmaiden; for, behold, from henceforth all generations shall call me blessed.

17. For Thou that art mighty hast done to me great things; and holy is Thy Name. And Thy mercy is on them that fear Thee from generation to generation.

18. Thou hast shewed strength with Thy arm; thou hast scattered the proud in the imagination of their hearts.

19. Thou hast put down the mighty from their seats and exalted the humble and the meek. Thou hast fill the hungry with good things and the rich Thou dost send empty away.

20. Thou dost help thy servant Israel, in remembrance of thy mercy: as Thou spakest to our ancestors to Abraham and to his seed for ever. And Mary abode with her about three months and returned to her own house.

21. And these are the words that Ioseph spake,, saying: Blessed be the God of our fathers and our mothers in Israel: for in an acceptable time Thou hast heard me, and in the day of salvation hast Thou helped me.

22. For Thou saids't I will preserve and make thee a covenant of the people to renew the face of the earth: and to cause the desolate places to be redeemed from the hands of the spoiler.

23. That thou mayest say to the captives, Go ye forth and be free; and to them that are in darkness, Show yourselves in the light. And they shall feed in the ways of pleasantness; and they shall no more hunt nor worry the creatures which I have made to rejoice before me.

24. They shall not hunger nor thirst any more neither shall the heat smite them nor the cold destroy them. And I will make on all My mountains a way for travellers; and My high places shall be exalted.

25. Sing ye heavens and rejoice thou earth; O ye deserts break forth with song: for Thou O God dost comfort Thy people; and console them that have suffered wrong.

Lection 2. 10.- "Ioseph begat (of Mary the Virgin, his wife) Iesus, who is called the Christ."-Curetonian and Lewis's Syriac, MS. ; and several of the oldest Latin MSS., in Matt. I.16, A.V.

Lection 2. 21-25.-The canticle of Ioseph here given is very similar to a certain portion of the book of Isaiah; indeed, appears to be taken from it, as Iohn borrowed from the Old Testament prophets. It has been omitted in all other Gospels extant. It is of singular beauty, and appropriate for use at Matins, as *Magnificat* is for Vesper, the Song of Zacharias finding an equally appropriate place at Nocturns.

Lection 3

The Nativity Of Iohn The Baptist

1. NOW Elisabeth's full time came that she should be delivered; and she brought forth a son. And her neighbours and her cousins heard how the Lord had shewed great mercy upon her; and they rejoiced with her.

2. And it came to pass, that on the eighth day they came to circumcise the child; and they called him Zacharias, after the name of his father. And his mother answered and said, Not so; but he shall be called Iohn. And they said unto her, There is none of thy kindred that is called by thy name.

3. And they made signs to his father, how he would have him called. And he asked for a writing table, and wrote, saying, his name is Iohn. And they all marvelled, for his mouth was opened immediately, and his tongue loosed, and he spake, and praised God.

4. And great awe came on all that dwelt round about them; and all these came on all that dwelt round about them; and all these sayings were made known abroad throughout all the hilly country of Judea. And all they that heard them laid them up in their hearts, saying, What manner of child shall this be! And the hand of Jova was with him.

5. And his father Zacharias was filled with the holy Spirit, and prophesied, saying, Blessed be thou, O God of Israel; for thou hast visited and redeemed thy people. And hast raised up an horn of salvation for us in the house of thy servant David. As thou spakest by the mouth of thy holy prophets, which have been since the world began.

6. That we should be saved from our enemies, and from the hand of all that hate us. To perform the mercy promised to our ancestors, and to remember thy holy covenant.

7. The oath which thou did'st sware to our father Abraham, that thou wouldest grant unto us, that we being delivered out of the hand of our enemies might serve thee without fear, in holiness and righteousness before thee all the days of our life.

8. And this child shalt be called the Prophet of the Highest: for he shalt go before Thy face, O God, to prepare Thy ways; to give knowledge of salvation unto Thy people by the remission of their sins.

9. Through the tender mercy of our God, whereby the dayspring from on high hath visited us; to give light to them that sit in darkness and in the shadow of death, to guide our feet into the way of peace.

10. And the child grew, and waxed strong in spirit, and his mission was hidden till the day of his shewing forth unto Israel.

Lection 4

Nativity of Iesus the Christ

1. NOW the birth of Iesu-Maria the Christ was on this wise. It came to pass in those days, that there went out a decree from Caesar Augustus, that all the world should be taxed. And all the people of Syria went to be taxed, every one into his own city, and it was midwinter.

2. And Ioseph with Mary also went up from Galilee, out of the city of Nazareth into Judea, unto the city of David, which is called Bethlehem (because they were of the house and lineage of David), to be taxed with Mary his espoused wife, who was great with child.

3. And so it was, that, while they were there, the days were accomplished that she should be delivered. And she brought forth her firstborn child in a Cave, and wrapped him in swaddling clothes, and laid him in a manger, which was in the cave; because there was no room for them in the inn. And behold it was filled with many lights, on either side Twelve, bright as the Sun in his glory.

4. And there were in the same cave an ox, and a horse, and an ass, and a sheep, and beneath the manger was a cat with her little ones, and there were doves also, overhead, and each had its mate after its kind, the male with the female.

5. Thus it came to pass that he was born in the midst of the animals which, through the redemption of man from ignorance and selfishness, he came to redeem from their sufferings, by the manifestation or the sons and the daughters of God.

6. And there were in the same country, shepherds abiding in the field, keeping watch over their flock by night. And when they came, lo, the angel of God came upon them, and the glory of the Highest shone round about them; and they were sore afraid.

7. And the angel said unto them, Fear not: for, behold, I bring you good tidings of great joy, which shall be to all people, for

unto you is born this day in the city of David a saviour, which is Christ, the Holy One of God. And this shall be a sign unto you; Ye shall find the babe wrapped in swaddling clothes lying in a manger.

8. And suddenly there was with the angel a multitude of the heavenly host praising God and saying, Glory to God in the highest, and on earth peace toward men of goodwill.

9. And it came to pass, as the angels were gone away from them into heaven, the shepherds said to one another, Let us now go even unto Bethlehem, and see this thing which is come to pass, which our God hath made known unto us.

10. And they came with haste, and found Mary and Ioseph in the cave, and the Babe lying in a manger. And when they had seen these things, they made known abroad the saying which was told them concerning the child.

11. And all they that heard it, wondered at those things told them by the shepherds; but Mary kept all these things, and pondered them in her heart. And the shepherds returned, glorifying and praising God for all the things that they had heard and seen.

12. AND when eight days were accomplished for the circumcising of the child, his name was called Iesu-Maria, as was spoken by the angel before he was conceived in the womb. And when the days of her purification according to the law of Moses were accomplished, they brought the child to Jerusalem, to present it unto God (as it is written in the law of Moses, every male that openeth the womb shall be called holy to the Lord).

13. And, behold, there was a man in Jerusalem, whose name was Simeon; and the same man was just and devout, waiting for the consolation of Israel; and the Holy Spirit was upon him. And it was revealed unto him that he should not see death, before he had seen the Christ of God.

14. And he came by the Spirit into the temple; and when the parents brought in the child Iesus, to do for him after the custom of the law, he perceived the child as it were a Pillar of light. Then took he him "up in his arms, and blessed God, and said:

15. Now lettest thou thy servant depart in peace, according to thy word. For mine eyes have seen thy salvation, which thou has prepared before the face of all people; to be a light to lighten the Gentiles, and to be the glory of thy people Israel. And his parents marvelled at those things which were spoken of him.

16. And Simeon blessed them, and said unto Mary his mother, Behold, this child is set for the falling and rising again of many in Israel; and for a Sign which shall be spoken against (yea, a sword shall pierce through thy own soul also), that the thoughts of many hearts may be revealed.

17. And there was one Anna, a prophetess, the daughter of Phanuel of the tribe of Aser, of a great age, who departed not from the temple, but served God with fastings and prayers night and day.

18. And she coming in that instant gave thanks likewise unto God, and spake of him to all them that looked for redemption in Jerusalem. And when they had performed all things according to the law they returned into Galilee, to their own city Nazareth.

Lection 4. 1 -The accepted date of the birth of Christ as corrected in the A. V. is A.M. 4000, or A.D. 1. This being so, his second visit to the Temple A.M. 4012, and after that his travels about A.M. 4018-4030; his Baptism A.M. 4031 ; His Transfiguration on the Mount, 4042 ; and his Crucifixion A.M. 4049, leaving eighteen years for his public ministry ; and his numerous teachings, which S. Iohn declares would fill a vast number of books, more than could be contained (comprehended by the world).

Lection 4. 4 - The animals here mentioned are sacred to the Deity in various countries and religions, the Cat and the Dove being specially honored and protected in Egypt (the most ancient centre of civilization, religion, philosophy and true science), as the symbols of Isis, the foreshadower of the "Divine Mother" of Christianity. Egypt (with her Trinity of Father, Mother, Child) gave refuge and sanctuary to the Infant Christ, Who came forth from thence to redeem humanity. The cat is not wilfully a "cruel animal," as falsely alleged by the ignorant, no more than the babe which torments it in ignorance of the pain it gives. Far more cruel are human beings, who torture and destroy millions of innocent creatures to gratify a depraved appetite or to minister to their vanity, or their lust for cruel experiment. The cat truly, as alleged by occultists, both ancient and modern, "the most human of all animals," and it is probable it was for this reason that it appears as the favourite animal of Iesus who was ever the friend of the despised, maligned and neglected although the most loving, gentle and graceful of all animals, rather than the more self assertive dog, especially as taught by man to hunt and to worry.

LECTION 4. 12.-Iesu Maria is the complete name. Iesus, he shall save, Maria, his people. Jesus is only the first part of the Holy Name, He saves His people, not at once, the entire human race, but those of goodwill- *homines boncæ voluntatis* - men and women of peace, and obedient to the divine law; and by these, their brethren through the ages, who *will* to be saved. The first part of the sacred Name seems to be generally used in the Gospel, as indicating that only the first part of his mission is *now*. When all men and women are gathered in, then will Christ be manifest as the complete Saviour, Iesu-Maria.

Lection 5

The Manifestation of Iesus to the Magi

1. Now when Iesus was born in Bethlehem of Judea, in the days of Herod the king, behold, there came certain Magi men from the east to Jerusalem, who had purified themselves and tasted not of flesh nor of strong drink, that they might find the Christ whom they sought. And they said, Where is he that is born King of the Jews? for we in the East have seen his Star, and are come to worship him.

2. When Herod the king had heard these things he was troubled, and all Jerusalem with him. And when he had gathered all the chief priests and scribes of the people together, he demanded of them where the Christ should be born.

3. And they said unto him, Bethlehem of Judea; for thus it is written by the prophet, and thou Bethlehem, in the land of Judea, art not the least among the princes of Judah; for out of thee shall come forth a Governor, that shall rule my people Israel.

4. Then Herod, when he had privily called the Magi, enquired of them diligently what time the Star appeared. And he sent them to Bethlehem, and said, Go and search diligently for the young child; and when ye have found him, bring me word again, that I may come and worship him also.

5. When they had heard the king, they departed; and, lo, the Star which the Magi of the East saw, and the angel of the Star went before them, till it came and stood over the place where the young child was, and the Star had the appearance of six rays.

6. And as they went on their way with their camels and asses laden with gifts, and were intent on the heavens seeking the child by the Star, they forgot for a little, their weary beasts who had borne thee burden and heat of the day, and were thirsty and fainting, and the Star was hidden from their sight.

7. In vain they stood and gazed, and looked one upon the other in their trouble. Then they bethought them of their camels and asses, and hastened to undo their burdens that they might have rest.

8. Now there was near Bethlehem a well by the way, And as they stooped down to draw water for their beasts, lo, the Star which they had lost appeared to them, being reflected in the stillness of the water.

9. And when they saw it they rejoiced with exceeding great joy.

10. And they praised God who had shewn his mercy unto them even as they shewed mercy unto their thirsty beasts.

11. And when they were come into the house, they saw the young child with Mary his mother, and fell down, and worshipped him: and when they had opened their treasures, they presented unto him gifts; gold, and frankincense, and myrrh.

12. And being warned of God in a dream that they should not return to Herod, they departed into their own country another way. And they kindled a fire according to their custom and worshipped God in the Flame.

13. And when they were departed, behold the angel of God appeared to Ioseph in a dream, saying, Arise, and take the young child and his mother, and flee into Egypt, and there remain until I bring thee word, for Herod will seek to destroy him.

14. AND when he arose, he took the young child and his mother by night, and departed into Egypt, and was there for about seven years until the death of Herod, that it might be fulfilled which was spoken of God by the prophet, saying, Out of Egypt have I called my son.

15. Elizabeth too when she heard it, took her infant son and went up into a mountain and hid him. And Herod sent his officers to Zacharias in the temple and said to him, Where is thy child? And he answered I am a minister of God and am continually in the temple. I know not where he is.

16. And he sent again, saying, Tell me truly where is thy son, Dost thou not know thy life is in my hand? And Zacharias answered, The Lord is witness if thou shed my blood, my spirit will God receive, for thou sheddest the blood of the innocent.

17. And they slew Zacharias in the Temple between the holy place and the altar; and the people knew it, for a voice was heard, Zacharias is slain, and his blood shall not be washed out until the avenger shall come. And after a time the priests cast lots, and the lot fell upon Simeon, and he filled his place.

18. Then Herod, when he saw that he was mocked of the wise men, was exceedingly wroth, and sent forth, and slew all the children that were in Bethlehem, and in all the coasts thereof, from two years old and under, according to the time which he had diligently enquired of the wise men.

19. Then was fulfilled that which was spoken by Jeremy the prophet, saying, In Rama was there a voice heard, lamentation, and weeping, and great mourning, Rachel weeping for her children, and would not be comforted, because they are not.

20. BUT when Herod was dead, behold, an angel of God appeared in a dream to Ioseph in Egypt. Saying, Arise, and take the young child and his mother, and return into the land of Israel: for they are dead which sought the young child's life.

21. And he arose, and took the young child and his mother and came into the land of Israel. And they came and dwelt in a city called Nazareth; and he was called the Nazarene.

LECTION 5. 9. -Note the beautiful lesson taught by these words. They look in vain for the signs of God who forget the needs of the poorer brethren and their beasts under their care. To look upon the needs of these who cannot speak (in human tongue) is to find the bright light they lose who only look upwards.

LECTION 5. 16. -Alluding to 2 Chron. xxiv. 20, in the Ierusalem Talmud, and also in the Babylonish, is an account of a priest named **Zacharias**, who was slain in the court of the priests near the altar, and whose blood never ceased to bubble from the earth, till a great number of priests and rabbins were slaughtered *(Talmud Hierosal,* fol. 69).

In the Protevangelium attributed to Iames, the first Bishop or Angel of the Church in Ierusalem is introduced the present story of Zacharias, and that Herod who slew the infants in Bethlehem slew also **Zacharias** the priest in the Temple when he said that he knew not where his infant son Iohn was hidden. It is this story, and not the incident in Chronicles, that most probably is referred to in a latter part of the Gospel by Iesus, being fresh in the memories of that generation, and so more likely to fasten attention.

Lection 6

The Childhood And Youth Of Iesus the Christ.
He Delivereth A Lion From The Hunters

1. NOW, Ioseph and Mary, his parents, went up to Jerusalem every year at the Feast of the Passover and they observed the feast after the manner of their brethren, who abstained from bloodshed and the eating of flesh and from strong drink. And when he was twelve years old, he went to Jerusalem with them after the custom of the feast.

2. And when they had fulfilled the days, as they returned, the child Iesus tarried behind in Jerusalem; and his parents knew not of it. But they, supposing him to have been in the company, went a day's Journey and they sought him among their kinsfolk and acquaintance. And when they found him not, turned back to Jerusalem, seeking him.

3. And it came to pass, that after three days they found him in the temple, sitting in the midst of the doctors, both hearing them, and asking them questions. And all that heard him were astonished at his understanding and answers.

4. And when they saw him, they were amazed; and his mother said unto him, Son, why hast thou thus dealt with us? Behold, thy father and I have sought thee sorrowing. And he said unto them, How is it that ye sought me? Wist ye not that I must be in my Parents' House. And they understood not the saying which he spake unto them. But his mother kept all these sayings in her heart.

5. And a certain prophet seeing him, said unto him, Behold the Love and the Wisdom of God are one in thee, therefore in the age to come shalt thou be called Iesu-Maria, for by the Christ shall God save mankind, which now is verily as the bitterness of the sea, but it shall yet be turned into sweetness, but to this generation the Bride shall not be manifest, nor yet to the age to come.

6. And he went down with them, and came to Nazareth, and was subject unto them. And he made wheels, and yokes, and tables also, with great skill. And Iesus increased in stature, and in favour with God and man.

7. AND on a certain day the child Iesus came to a place where a snare was set for birds, and there were some boys there. And Iesus said to them, who hath set this snare for the innocent creatures of God? Behold in a snare shall they in like manner be caught. And he beheld twelve sparrows as it were dead.

8. And he moved his hands over them, and said to them, Go, fly away, and while ye live remember me. And they arose and fled away making a noise. And the Jews, seeing this, were astonished and told it unto the priests.

9. And other wonders did the child, and flowers were seen to spring up beneath his feet, where there had been naught but barren ground before. And his companions stood in awe of him.

10. AND in the eighteenth year of his age, Iesus was espoused unto Miriam, a virgin of the tribe of Judah with whom he lived seven years, and she died, for God took her, that he might go on to the higher things which he had to do, and to suffer for the sons and daughters of men.

11. And Iesus, after that he had finished his study of the law, went down again into Egypt that he might learn of the wisdom of the Egyptians, even as Moses did. And going into the desert, he meditated and fasted and prayed, and obtained the power of the Holy Name, by which he wrought many miracles.

12. And for seven years he conversed with God face to face, and he learned the language of birds and of beasts, and the healing powers of trees, and of herbs, and of flowers, and the hidden secrets of precious stones, and he learned the motions of the Sun and the Moon and the stars, and the powers of the letters, and mysteries of the Square and the Circle and the Transmutation of things, and of forms, and of numbers, and of signs. From thence he returned to Nazareth to visit his parents, and he taught there and in Jerusalem as an accepted Rabbi, even in the temple, none hindering him.

13. AND after a time he went into Assyria and India and into Persia and into the land of the Chaldeans. And he visited their temples and conversed with their priests, and their wise men for many years, doing many wonderful works, healing the sick as he passed through their countries.

14. And the beasts of the field had respect unto him and the birds of the air were in no fear of him, for he made them not afraid, yea even the wild beasts of the desert perceived the power of God in him, and did him service bearing him from place to place.

15. For the Spirit of Divine Humanity filling him, filled all things around him, and made all things subject unto him, and thus shall yet be fulfilled the words of the prophets, The lion shall lie down with the calf, and the leopard with the kid, and the wolf with the lamb, and the bear with the ass, and the with the dove. And a child shall lead them.

16. And none shall hurt or destroy in my holy mountain, for the earth shall be full of the knowledge of the Holy One even as the waters cover the bed of the sea. And in that day I will make again a covenant with the beasts of the earth and the fowls of the air, and the fishes of the sea and with all created things. And will break the bow and the sword and all the instruments of warfare will I banish from the earth, and will make them to lie down in safety, and to live without fear.

17. And I will betroth thee unto me for ever in righteousness and in peace and in loving kindness, and thou shalt know thy God, and the earth shalt bring forth the corn the wine and the oil, and I will say unto them which were not my people, Thou art my people; and they shall say unto me, Thou art our God.

18. And on a certain day as he was passing by a mountain side nigh unto the desert, there met him a lion and many men were pursuing him with stones and javelins to slay him.

19. But Iesus rebuked them, saying, Why hunt ye these creatures of God, which are more noble than you? By the cruelties of many generations they were made the enemies of man who should have been his friends.

20. If the power of God is shown in them, so also is shown his long suffering and compassion. Cease ye to persecute this creature who desireth not to harm you, see ye not how he fleeth from you, and is terrified by your violence?

21. And the lion came and lay at the feet of Iesus, and shewed love to him; and the people were astonish , and said, Lo, this man loveth all creatures and hath power to command even these beasts from the desert, and they obey him.

LECTION 6. 5. -In what way this prediction is to be fulfilled is not as yet made manifest - whether Iesus shall yet be manifest and received by his people as the Two-in-One, the All-gentle as well as the All-powerful, or whether He shall assume the feminine form, or whether He shall be manifest with His counterpart. Many false Christs shall come with signs and lying wonders.

LECTION 6. 10. -Iosephus mentions a section of the Essenes, or Iessenes, who, unlike the great majority of them, lived in "honourable marriage," observing their rules and customs in all other matters, such as abstinence from blood sacrifices, flesh eating, etc. Some consider it most probable, therefore, that at this period Iesus married, according to the usual custom of the Iews, and in his case especially, that he might have full experience of human life, and thus be a perfect Example for all, knowing the joys and sorrows of all,-and that it was just before his further travels preparatory to his entrance into the Ministry that he lost by death the wedded partner of his youth. He was " in all things like as we are, yet without sin"

Lection 7

The Preaching Of Iohn The Baptist

1. NOW in the fifteenth year of the reign of Tiberius Caesar, Pontius Pilate being governor of Judea, and Herod being tetrarch of Galilee (Caiaphas being the high priest, and Annas chief of the Sanhedrim) the word of God came unto Iohn the son of Zacharias, in the wilderness.

2. And he came into all the country about Jordan, preaching the baptism of repentance for the remission of sins. As it is written in the prophets, Behold I send my messenger before thy face, who shall prepare thy way before thee; the voice of one crying in the wilderness, Prepare ye the way of the Holy One, make straight the paths of the Anointed.

3. Every valley shall be filled, and every mountain and hill shall be brought low; and the crooked shall be made straight, and the rough ways shall be made smooth. And all flesh shall see the salvation of God.

4. And the same Iohn had his raiment of camel's hair, and a girdle of the same about his loins, and his meat was the fruit of the locust tree and wild honey. Then went out to him Jerusalem, and all Judea, and all the region round about Jordan, and were baptized of him in the Jordan confessing their sins.

5. And he said to the multitude that came forth to be baptized of him, O generation of disobedient ones, who hath warned you to flee from the wrath to come? Bring forth therefore fruits worthy of repentance and begin not to say within yourselves, We have Abraham to our father.

6. For I say unto you, that God is able of these stones to raise up children unto Abraham. And now also the axe is laid unto the root of the trees: every tree therefore which bringeth not forth good fruit is hewn down, and cast into the fire.

7. And the wealthier people asked him, saying, What shall we do then? He answereth and saith unto them, He that hath two coats, let him impart to him that hath none; and he that hath food let him do likewise.

8. Then came also certain taxgatherers to be baptised and said unto him, Master, what shall we do? And he said unto them, Exact no more than that which is appointed you, and be merciful after your power.

9. And the soldiers likewise demanded of him, saying, And what shall we do? And he said unto them, Do violence to no man, neither accuse any falsely; and be content with sufficient wages.

10. And to all he spake, saying, Keep yourselves from blood and things strangled and from dead bodies of birds and beasts, and from all deeds of cruelty, and from all that is gotten of wrong; Think ye the blood of beasts and birds will wash away sin! I tell you Nay, Speak the Truth. Be just, Be merciful to one another and to all creatures that live, and walk humbly with your God.

11. And as the people were in expectation, and all men mused in their hearts of Iohn, whether he were the Christ or not, Iohn answered; saying unto them all, I indeed baptize you with water; but One mightier than I cometh, the latchet of whose shoes I am not worthy to unloose.

12. He shall also baptize you with water and with fire. Whose fan is in his hand, and he will thoroughly purge his floor, and will gather the wheat into his garner; but the chaff he will burn with fire unquenchable. And many other things in his exhortation preached he unto the people.

LECTION 7. 4.-The fruit of the Carob tree ("S. Iohn's Bread") ; not the insect of that name, as is supposed by the people in general.

LECTION 7. 10,-As noticed before, the Essenes did not frequent the blood sacrifices of the Temple. Iohn and Iesus acted accordingly.

Lection 8

The Baptism of Iesu Maria The Christ

1. AND it was in the midst of the summer, the tenth month. Then cometh Iesus from Galilee to Jordan unto Iohn, to be baptized of him. But Iohn forbade him, saying, I have need to be baptized of thee, and comest thou to me? And Iesus answering said unto him, Suffer it to be so now, for thus it becometh us to fulfil all righteousness. Then he suffered him.

2. And Iesus, when he was baptized, went up straightway out of the water; and, lo, the heavens were opened unto him, and a bright cloud stood over him, and from behind the cloud Twelve Rays of light, and thence in the form of a Dove, the Spirit of God descending and lighting upon him. And, lo, a voice from heaven saying, This is my beloved Son, in whom I am well pleased; this day have I begotten thee.

3. And Iohn bare witness of him, saying, This was he of whom I spake, He that cometh after me is preferred before me, for he was before me. And of his fulness have all we received, and grace for grace. For the law was in part given by Moses, but grace and truth cometh in fulness by Iesus Christ.

4. No man hath seen God at any time. The only begotten which cometh from the bosom of the Eternal in the same is God revealed. And this is the record of Iohn, when the Jews sent priests and Levites from Jerusalem to ask him, Who art I thou? And he deified not, but confessed I am not the Christ.

5. And they asked him, What then? Art thou Elias? And he saith, I am not, Art thou that prophet of whom Moses spake? And he answered, No. Then said they unto him, Who art thou? that we may give an answer to them that sent us. What sayest thou of thyself? And he said, I am the voice of one crying in the wilderness, Make straight the way of the Holy One, as said the Prophet Esaias.

6. And they which were sent were of the Pharisees, and they asked him and said unto him, Why baptizest thou then, if thou be not that Christ, nor Elias, neither that prophet of whom Moses spake?

7. Iohn answered them, saying, I baptize with water; but there standeth One among you, whom ye know not, He shall baptize with water and with fire. He it is who coming after me is preferred before me, whose shoe's latchet I am not worthy to unloose.

8. These things were done in Bethabara, beyond Jordan, where Iohn was baptizing. And Iesus began at this time to be thirty years of age, being after the flesh indeed the Son of Ioseph and Mary; but after the Spirit. the Christ, the Son of God, the Father and Mother Eternal, as was declared by the Spirit of holiness with power.

9. AND Ioseph was the son of Jacob and Elisheba, and Mary was the daughter of Eli (called Joachim) and Anna, who were the children of David and Bathsheba, of Judah and Shela, of Jacob and Leah, of Isaac and Rebecca, of Abraham and Sarah, of Seth and Maat, of Adam and Eve, who were the children of God.

LECTION 8. 2. -This "bright light" at his baptism is mentioned in the "Gospel of the Hebrews," which is undoubtedly the original Gospel of S. Matthew, and the one used in the primitive Church of Ierusalem, and identical with this.

Iustin Martyn quotes this Gospel as the original Gospel of Matthew, and endeavours to explain away the supposed "heresy" in the words, "**This day have I begotten thee**," which shows that the present Gospel of Matthew could not have been extant in his time, else he would have quoted it with gladness as omitting these words.

v. 7.-The earthly ministry of Iesus, beginning at thirty years of age, complete and continuing till his death at the age of forty-nine, must therefore have lasted much longer than is generally supposed, even eighteen years. During the latter part of it, the Iews who knew him attested that he was then " not fifty years old."

Lection 9

The Four Temptations

1. THEN was Iesus led up of the spirit into the wilderness to be tempted of the Devil. And the wild beasts of the desert were around him, and became subject unto him. And when he had fasted forty days and forty nights he was afterwards an hungered.

2. And when the tempter came to him, he said, If thou be the Son of God, command that these stones be made bread, for it is written, I will feed thee with the finest of wheat and with honey, out of the rock will I satisfy thee.

3. But he answered and said, It is written, Man shall not live by bread alone, but by every word that proceeded out of the mouth of God.

4. Then the Devil placeth before him a woman, of exceeding beauty and comeliness and of subtle wit, and a ready understanding withal, and he said unto him. Take her as thou wilt, for her desire is unto thee, and thou shalt have love and happiness and comfort all thy life, and see thy children's children, yea is it not written, It is not good for man that he should be alone?

5. And Iesu-Maria said, Get thee behind me, for it is written, Be not led away by the beauty of woman, yea, all flesh is as grass and the flower of the field; the grass withereth and the flower fadeth away, but the Word of the Eternal endureth for ever. My work is to teach and to heal the children of men, and he that is born of God keepeth his seed within him.

6. And the Devil taketh him up into the holy city, and setteth him on a pinnacle of the Temple. And saith unto him, If thou be the Son of God, cast thyself down; for it is written, He shall give his angels charge concerning thee; and in their hands they shall bear thee up lest at any time thou dash thy foot against a stone.

7. And Iesus said unto him, It is written again, Thou shalt not tempt the Lord thy God.

8. Then the Devil took him up into an exceeding high mountain in the midst of a great plain and, round about, twelve cities and their peoples, and shown from thence he shown unto him all the kingdoms of the world in a moment of time. And the Devil said unto him, All this power will I give thee, and the glory of them: for that is delivered unto me; and to whomsoever I will, I give it: for it is written, thou shalt have dominion from sea to sea, so thou shalt judge thy people with righteousness and thy poor

with mercy, and. make a full end of oppression. If thou therefore wilt worship me, all shall be thine.

9. And Iesu-Maria answered and said unto him, get thee behind me, Satan; for it is written, Thou shalt worship thy God, and Him only shalt thou serve. Without the power of God, the end of evil cannot come.

10. Then the Devil having ended all the temptations leaveth him and departed for a season. And behold, angels of God came and ministered unto him.

LECTION 9. 1 -The Essenes or Nazarenes, somewhat like the Indian Yogi, sought to attain divine union by solitary meditation in unfrequented places. In the monastery of our Lord on the summit of Quarantania, a cell is shown with rude frescoes of the event. This mountain is about 18,000 feet high, in a barren and desolate region east of Ierusalem, north of the road to Iericho, overlooking the valley of the Iordan.

v. 2-9 -Observe, the temptations are addressed to the fourfold nature of man, as recognised by the ancient Egyptians. 1st.-To the outer body, with its physical needs. 2nd. -To the inner body, the seat of the senses and desires. 3rd.-To the soul, the seat of the intellect.

LECTION 9. 3 -In all the ancient initiations woman was one of the temptations placed in the way of the aspirant. That this was not omitted in the trial of the "Perfect Man" we may be certain, and we are expressly told in the Epistle to the Hebrews that "he was in all points tempted even as we are." Why the writers of the Canonical Gospels omitted this trial, or whether it was dropped out of the original by accident we cannot say, but here we have it restored in its place. It is evidently inculcated by Iesus in this second temptation (what has always been known to the wise) that adepts should store up their physical strength for work on a higher plane, and this Iesus did for the work of the ministry as an example for all who would follow him and heal the bodies and souls of others.

Here we have one of the many passages which show that the words attributed to the writers of the Epistles are quotations from this Gospel, and that such portions at least were extant in their time.-e.g., I. Iohn iii. 9. (A. V.).

Lection 10

Ioseph And Mary Make A Feast Unto Iesus.
Andrew And Peter Find Iesus.

1. AND when he had returned from the wilderness, the same day, his parents made him a feast, and they gave unto him the gifts which the Magi had presented to him in his infancy. And Mary said, These things have we kept for thee even to this day, and she gave unto him the gold and the frankincense and the myrrh. And he took of the frankincense, but of the gold he gave unto his parents for the poor, and of the myrrh he gave unto Mary who is called Magdalene.

2. Now this Mary was of the city of Magdala in Galilee. And she was a great sinner, and had seduced many by her beauty and comeliness. And the same came unto Iesus by night and confessed her sins, and he put forth his hand and healed her, and cast out of her seven demons, and he said unto her, Go in peace, thy sins are forgiven thee. And she arose and left all and followed him, and ministered unto him of her substance, during the days of his ministry in Israel.

3. THE next day Iohn saw Iesus coming unto him, and said, Behold the Lamb of God, which by righteousness taketh away the sin of the world. This is he of whom I said, He was before me; and I knew him not; but that he should be made manifest to Israel; therefore am I come baptizing with water.

4. And Iohn bare record, saying, I saw the Spirit descending from heaven like a Dove, and it abode upon him. And I knew him not, but he that sent me to baptize with water, the same said unto me, Upon whom thou shalt see the Spirit descending, and remaining on him, the same is he which baptized with water and with fire, even the Spirit. And I saw, and bare record that this was the Son of God.

5. THE day after, Iohn stood by the Jordan and two of his disciples. And looking upon Iesus as he walked, he saith, Behold the Christ, the Lamb of God! And the two disciples heard him speak, and they followed Iesus.

6. Then Iesus turned and saw them following and saith unto them, What seek ye? They said unto him, Rabbi (which is, being interpreted, Master), where dwellest thou? He saith unto them, Come and see. They came and saw where he dwelt, and abode with him that day: for it was about the tenth hour.

7. One of the two which heard Iohn speak and followed him was Andrew, Simon Peter's brother. He first findeth his own brother Simon and said unto him, We have found the Messias, which is, being interpreted the Christ. And he brought him to Iesus And when Iesus beheld him, he said, Thou art Simon Bar Jona: thou shalt be called Kephas (which is, by interpretation, a rock).

8. THE day following, Iesus goeth forth into Galilee, and findeth Philip, and saith unto him, Follow me. Now Philip was of Bethsaida, the city of Andrew and Peter. Philip findeth Nathanael, who is called Bar Tholmai, and saith unto him, We have found him, Of whom Moses in the law and the Prophets did write, Iesus of Nazareth, the son of Ioseph and Mary, And Nathanael said unto him, Can there any good thing come out of Nazareth ? Philip said unto him, Come and see.

9. Iesus saw Nathanael coming to him and saith of him, Behold an Israelite indeed, in whom is no guile! Nathanael saith unto him, Whence knowest thou me? Iesus answered and said unto him, Before that Philip called thee, when thou wast under the Fig tree, I saw thee. Nathanael answered and saith unto him, Rabbi, thou art the Son of God. thou art the King of Israel. Yea, under the Fig tree did I find thee.

10. Iesus answered and said unto him, Nathanael Bar Tholmai, because I said unto thee, I saw thee under the Fig tree, believest thou ? thou shalt see greater things than these. And he saith unto him, Verily, verily, I say unto you, hereafter ye shall see heaven open, and the angels of God ascending and descending upon the Son of man.

The Gospel of the Holy Twelve
Translated from the original Aramaic
by Rev. G.J.R. Ouseley

Section 2, Lections 11 thru 20

Lection 11

The Anointing By Mary Magdalene

1. AND one of the Pharisees desired him that he would eat with him. And he went into the Pharisee's house and sat down to eat.

2. And behold a certain woman of Magdala, who was reputed to be a sinner, was in the city, and when she knew that Iesus sat at meat in the Pharisee's house, she brought an Alabaster box of ointment, and stood at his feet behind him, weeping, and washed His feet with tears, and did wipe them with the hairs of her head and kissed his feet, and anointed them with ointment.

3. Now when the Pharisee which had bidden him saw it, he thought within himself, saying, This man, if he were a prophet, would have known who and what manner of woman this is that toucheth him: for she is a sinner.

4. And Iesus answering said unto him, Simon, I have somewhat to say unto thee. And he saith, Master, say on.

5. There was a certain creditor which had two debtors: the one owed five hundred pence and the other fifty. And when they had nothing to pay, he frankly forgave them both. Tell me, therefore, which of them will love him most.

6. Simon answered and said, I suppose that he to whom he forgave most. And he said unto him, Thou hast rightly judged.

7. And he said unto Simon, Seest thou this woman? I entered into thine house, thou gavest me no water for my feet; but she hath washed my feet with tears and wiped them with the hairs of her head. Thou gavest me no kiss: but this woman since the time I came in hath not ceased to kiss my feet. My head with oil thou didst not anoint: but this woman hath anointed my feet with ointment.

8. Wherefore I say unto thee, man but also beast and birds of the air, yea, even the fishes of the sea; but to whom little is forgiven, the same loveth little. Her sins, which are many, are forgiven, for she loved much, not only man but also beast and birds of the air, yea, even the fishes of the sea; but to whom little is forgiven, the same loveth little.

9. And he said unto her, Thy sins are forgiven, and they who sat at the table began to say within themselves, who is this that forgiveth sins also?

10. Though he had said not, I forgive thee, but Thy sins are forgiven thee, for he discerned true faith and penitence in her heart. And Iesus needed not that any should testify of any man, for he himself knew what was in man.

LECTION 11. 1-2.-There are two anointings by Mary Magdalene recorded. The first was to his prophetical ministry, the last preparatory to his self-oblation unto death on the cross in the upper room, and his subsequent murder by the Roman authorities and the Iewish priests.

Lection 12

The Marriage In Cana
The Healing of the Nobleman's Son

1. AND the next day there was a marriage in Cana of Galilee; and the mother of Iesus was there: And both Iesus and Mary Magdalene were there, and his disciples came to the marriage.

2. And when they wanted wine the mother of Iesus saith unto him, They have no wine. Iesus saith unto her, Woman, what is that to thee and to me ? mine hour is not yet come. His mother saith unto the servants, Whatsoever he saith unto you, do it.

3. And there were set there six waterpots of stone, after the manner of the purifying of the Jews, containing two or three firkins apiece. And Iesus saith unto them, Fill the waterpots with water. And they filled them up to the brim. And he said unto them, Draw out now, and bear unto the governor of the feast. And they bare it.

4. When the ruler of the feast had tasted the water that was made wine to them, and knew not whence it was; the governor of the feast called the bridegroom, and saith unto him. Every man at the beginning doth set forth good wine and when men have well drunk, then that which is worse; but thou hast kept the good wine until now.

5. This beginning of miracles did Iesus in Cana of Galilee, and manifested forth his glory; and many disciples believed on him.

6. After this he went down to Capernaum, he, and his mother, with Mary Magdalene, and his brethren, and his disciples: and they continued there for many.

7. And there arose a question between some of John's disciples and the Jews about purifying. And they came unto John, and said unto him, Rabbi, he that was with thee beyond Jordan, to whom thou bearest wittness, behold, the same baptizeth, and all do come to him.

8. John answered and said, A man can receive nothing, except it be given him from heaven. Ye yourselves bear me witness, that I said, I am not the Christ, but that I am sent before him.

9. He that hath the bride is the bridegroom; but the friend of the bridegroom, which standeth and heareth him, rejoiceth greatly because of the bridegroom's voice; this my joy therefore is fulfilled. He must increase; but I must decrease. He that is of the earth is earthly, and speaketh of the earth: he that cometh from heaven is above all.

10. AND certain of the Pharisees came and questioned Iesus, and said unto him, how sayest thou that God will condemn the world ? And Iesus answered, saying, God so loveth the world, that the only begotten Son is given, and cometh into the world, that whosoever believeth in him may not perish, but have everlasting life. God sendeth not the Son into the world to condemn the world; but that the world through him may be saved.

11. They who believe on him are not condemned: but they that believe not are condemned already, because they have not believed in the name of the only begotten of God. And this is the condemnation, that the light is come into the world, and men love darkness rather than light, because their deeds are evil .

12. For all they that do evil hate the light, neither come they to the light, lest their deeds may be condemned. But they that do righteousness come to the light, that their deeds may be made manifest, that they are wrought in God.

13. AND there was a certain nobleman, whose son was sick at Capernaum. When he heard that Iesus was come into Galilee, he went unto him, and besought him that he would come down, and heal his son; for he was at the point of death.

14. Then said Iesus unto him, Except ye see signs and wonders, ye will not believe. The nobleman saith unto him, Sir, come down ere my child die.

15. Iesus saith unto him, Go thy way; thy son liveth. And the man believed the word that Iesus had spoken unto him, and he went his way. And as, he was now going down, his servants met him, and told him, saying, Thy son liveth.

16. Then enquired he of them the hour when he began to amend. And they said unto him, Yesterday of the seventh hour the fever left him. So the father knew that it was at the same hour, in the which Iesus said unto him, Thy son liveth. And himself believed, and his whole house.

LECTION 12. 3-4.-Iesus being a Yessene (Essene) could not drink intoxicating wine, and it is to be remarked here, that he did not provide it. He poured water into jars, and they tasted it as wine unfermented, or, if fermented, with four times or least twice its volume of water, which makes what is termed all through the Gospel the "*fruit of the vine.*" It is impossible that Iesus could sanction drunkenness, though his enemies slandered him as a "wine-bibber."

v. 16.-Two modes of reckoning time were in use. The Roman, from 12 midnight to 12 midnight. The Iewish from 6 a.m. (mean time is here spoken of) in the even to 6 p.m. of next even. The Iewish hours, adopted from the Temple in the Christian Church in her devotions, were as follows:

6 p.m. 1st watch, Vespers. Ferial.
9 p.m. 2nd watch. Nightfall ("Compline" Lat. use).
12 midnight 3rd watch. Nocturms.
3 a.m. 4th watch. Daybreak. (Lauds).
5-6 p.m. Seventh or last hour of the night.
6 a.m. Matins (or "Prime" Lat. use). First hour.
9 a.m. Terce. Third hour.
12 midday. Sext. Sixth hour.
3 p.m. Nones. Ninth hour.
5-6 p.m. Eleventh or last hour of the day. Vespers. Festal. Really "Compline" in its true sense.
The "seventh hour," in this place is therefore 1 p.m. of our reckoning; whether by Iewish or Roman time- 13th hour in some countries.

Lection 13

The First Sermon In The Synagogue Of Nazareth

1. AND Iesus came to Nazareth, where he had been brought up: and, as his custom was, he went into the synagogue on the sabbath day, and stood up for to read. And there was delivered unto him the roll of the prophet Esaias.

2. And when he had opened the roll, he found the place where it was written. The Spirit of the Lord Is upon me, because he hath anointed me to preach the gospel to the poor; he hath sent me to heal the brokenhearted, to preach deliverance to the captives and recovering of sight to the blind, to set at liberty them that are bound. To preach the acceptable year of the Lord.

3. And he closed the roll, and gave it again to the minister, and sat down, And the eyes of all them that were in the synagogue were fastened on him. And he began saying unto them. This day is this scripture fulfilled in your ears. And all bare him witness, and wondered at the gracious words which proceeded out of his mouth. And they said, Is not this Ioseph's son?

4. And some brought unto him a blind man to test his power, and said, Rabbi, here is a son of Abraham blind from birth. Heal him as thou hast healed Gentiles in Egypt. And he, looking upon him, perceived his unbelief and the unbelief of those that brought him, and their desire to ensnare him. And he could do no mighty work in that place because of their unbelief.

5. And they said unto him, Whatsoever we have heard done in Egypt, do also here in thy own country. And he said, Verily I say unto you, No prophet is accepted in his own home or in his own country, neither doth a physician work cures upon them that know him.

6. And I tell you of a truth, many widows were in Israel in the days of Elias, when the heaven was shut up three years and six months, when great famine was throughout all the land. But unto none of them was Elias sent, save unto Sarepta, a city of Sidon, unto a woman that was a widow.

7. And many lepers were in Israel in the time of Eliseus the prophet; and none of them was cleansed, saving Naaman the Syrian.

8. And all they in the synagogue, when they heard these things, were filled with wrath. And rose up, and thrust him out of the city, and led him unto the brow of the hill whereon their city was built, that they might cast him down headlong. But he, passing through the midst of them, went his way and escaped them.

LECTION 13. 5. -The effects of his education in Egypt and his travels in other countries and knowledge of their religion and mysteries are here clearly seen in the largeness of the heart of Iesus, and his sympathy with all men. He is the true Catholic, who excludes none from his love whose hearts are unto righteousness, while he pities those that are not, knowing the terrible fate that awaits them.

Lection 14

The Calling Of Andrew And Peter
The Teaching of Cruelty in Animals
The Two Rich Men

1. NOW Herod the tetrarch, being reproved by John the Baptist for Herodias his brother Philip's wife, and for all the evils which he had done, added yet this above all, that he shut up John in prison.

2. And Iesus began to preach, and to say, Repent; for the kingdom of heaven is at hand. And as he was walking by the sea of Galilee, he saw Simon called Peter, and Andrew his brother, casting a net in the sea; for they were fishers. And he saith unto them, Follow me, and I will make you fishers of men. And they straightway forsook their nets, and followed him.

3. And going on from thence, he saw other two brethren, James the son of Zebedee, and John his brother, in a ship with Zebedee their father, mending their nets; and he called them. And they immediately left their nets, and the ship, and their father, and followed him.

4. And Iesus went about all Galilee, teaching in, their synagogues, and preaching the gospel of the kingdom, and healing all manner of sickness and all manner of disease among the people. And the fame of his miracles went throughout all Syria, and they brought unto him many sick people that were taken with divers diseases and torments, and those which were lunatick, and those that had the palsy, and he healed them.

5. And there followed him great multitudes of people from Galilee, and from Decapolis, and from Jerusalem, and from Judea, and from beyond Jordan.

6. AND as Iesus was going with some of his disciples he met with a certain man who trained dogs to hunt other creatures. And he said to the man, Why doest thou thus? and the man said, By this I live and what profit is there to any in these creatures? these creatures are weak, but the dogs they are strong. And Iesus said, Thou lackest wisdom and love. Lo, every creature which God hath made hath its end, and purpose, and who can say what good is there in it? or what profit to thyself, or mankind?

7. And, for thy living, behold the fields yielding their increase, and the fruit-bearing trees and the herbs; what needest thou more than these which honest work of thy hands will not give to thee? Woe to the strong who misuse their strength, Woe to the hunters for they shall be hunted.

8. And the man marvelled, and left off training the dogs to hunt, and taught them to save life rather than destroy, And he learned of the doctrines of Iesus and became his disciple.

9* AND behold there came to him two rich men, and one said, Good Master. But he said, Call me not good, for One alone is the All good, and that is God.

10. And the other said to him, Master, what good thing shall I do and live? Iesus said, Perform the Law and the prophets. He answered, I have performed them. Iesus answered, Go, sell all thou hast and divide with the poor, and follow me. But this saying pleased him not.

11. And the Lord said unto him, How sayest thou that thou hast performed the Law and the prophets? Behold many of thy brethren are clad with filthy rags, dying from hunger and thy house is full of much goods, and there goeth from it nought unto them.

12. And he said unto Simon, It is hard for the rich to enter the kingdom of heaven, for the rich care for themselves, and despise them that have not.

LECTION 14. 4. -Miracles are not violations of the laws of nature, but rather suspensions of lower by higher laws- wonders wrought by using wisely the subtle forces of nature, (whether by seen or unseen agencies) unknown to the science of the day, and in advance of the knowledge of the people. Many are the spiritual agencies, the knowledge of which we are now recovering, but which have existed and acted all through the ages. Occult phenomena also appear to have been used by religious teachers in all ages in the East, to attract the attention of the listless and thoughtless, and having roused and secured their interest, to teach them spiritual truths or give them higher revelations; just as in the West. in modern times, they Bound a bell, or sing an " invitatory" or a hymn to "call the people."

Lection 15

Healing Of The Leper And The Man With Palsy
The Deaf Man who Denied that Others could Hear

1. AND it came to pass, when he was in a certain city, behold a man full of leprosy, who, seeing Iesus, fell toward the earth, and besought him, saying, Lord if thou wilt, thou canst make me clean. And he put forth his hand, and touched him, saying, Blessed be thou who believest; I will, be thou clean. And immediately the leprosy departed from him.

2. And he charged him saying, Tell no man: but go, and shew thyself to the priest, and offer for thy cleansing, according as Moses commanded, for a testimony unto them. But so much the more went there a fame abroad of him; and great multitudes came together to hear, and to be healed by him of their infirmities. And he withdrew himself into the wilderness, and prayed.

3. AND it came to pass on a certain day, as he was teaching, that there were Pharisees and doctors of the law sitting by, to see them which were come out of every town, of Galilee, and Judea, and Jerusalem, and the power of God was present to heal them.

4. AND, behold, they brought in a bed a man who was taken with a palsy: and they sought means to bring him in, and to lay him before him. And when they could not find by what way they might bring him in because of the multitude, they went upon the housetop, and let him down through the tiling with his couch into the midst before Iesus. And when he saw their faith, he said unto him, Man, thy sins are forgiven thee.

5. And the scribes and the pharisees began to reason, saying, Who is this which speaketh blasphemies? Who can forgive sins, but God alone? But when Iesus perceived their thoughts, he answering said unto them, What reason ye in your hearts? Can even God forgive sins, if man repent not? Who said, I forgive thee thy sins? Said I not rather, Thy sins are forgiven thee?

6. Whether is easier to say. Thy sins be forgiven thee; or to say, Rise up and walk? But that ye may know that the Son of Man hath power upon earth to discern, and declare the forgiveness of sins (he said unto the sick of the palsy), I say unto thee, Arise, and take up thy couch, and go to thine house.

7. And immediately he arose up before them, and took up that whereon he lay, and departed to his own house, glorifying God. And they were all amazed, and they glorified God, and were filled with the Spirit of reverence, saying, We have seen strange things to day.

8. AND as Iesus was going into a certain village there met him a man who was deaf from his birth. And he believed not in the sound of the rushing wind, or the thunder, or the cries of the beasts, or the birds which complained of their hunger or their hurt, nor that others heard them.

9. And Iesus breathed into his ears, and they were opened, and he heard. And he rejoiced with exceeding joy in the sounds he before denied. And he said, Now hear all things.

10. But Iesus said unto him. How sayest thou, I hear all things? Canst thou hear the sighing of the prisoner, or the language of the birds or the beasts when they commune with each other, or the voice of angels and spirits? Think how much thou canst not hear, and be humble in thy lack of knowledge.

LECTION 15. 4. -The houses in Palestine were constructed with flat roofs, and entrance was easily made into the court below without entering by the door below.

Lection 16

Calling of Matthew
Parable of the New Wine in the Old Bottles

1. AND after these things he went forth, and saw a tax gatherer, named Levi, sitting at the receipt of custom: and he said unto him, Follow me. And he left all, rose up, and followed him.

2. And Levi made him a great feast in his own house: and there was a great company of taxgatherers and of others that sat down with them. But the Scribes and Pharisees murmured against his disciples, saying, Why do ye eat and drink with publicans and sinners ?

3. And Iesus answering said unto them, They that are whole need not a physician; but they that are sick. I came not to call the righteous, but sinners to repentance.

4. And they said unto him, Why do the disciples of John fast often, and make prayers, and likewise the disciples of the Pharisees; but thine do eat and drink ?

5. And he said unto them, Wherewith shall I liken the men of this generation, and to what are they like? They are like unto children, sitting in the market place and calling one to another and saying, We have piped unto you, and ye have not danced, we have mourned to you and ye have not lamented.

6. For John the Baptist came neither eating nor drinking, and ye say, He hath a devil, The Son of Man cometh eating and drinking the fruits of the earth, and the milk of the flock, and the fruit of the vine, and ye say, Behold a glutton and wine bibber, a friend of publicans and sinners.

7. Can ye make the children of the bridechamber fast, while the bridegroom is with them? But the days will come, when the bridegroom shall be taken away from them, and then shall they fast in those days.

8. AND he spake also this parable unto them, saying, No man putteth a piece of new cloth upon an old garment; for then the new agreeth not with the old, and the garment is made worse.

9. And no one putteth new wine into old bottles; else the new wine will burst the bottles, and be spilled, and the bottles shall perish. But new wine must be put into new bottles, and both are preserved.

10. None also having drunk old wine, straightway desire new: for they say, The old is better. But the time cometh when the new shall wax old, and then the new shall be desired by them. For as one changeth old garments for new ones, so do they also change the body of death for the body of life, and that which is past for that which is coming.

LECTION 16. 1. -"Levi" is by tradition identified with Matthew, the writer of the second of the four Gospels (as received by the Church), Mark being the first of the four Evangelists, though placed second in the A. V.

v. 9. -It was the custom in Palestine to use the skins of animals to hold wine as we do glass bottles, and such leathern bottles when filled with new wine were liable to burst by reason of the fermentation of the wine within them.

Lection 17

Iesus Sendeth Forth The Twelve and their Fellows

1. AND Iesus went up into a mountain to pray. And when he had called unto him his twelve disciples, he gave them power against unclean spirits to cast them out and to heal all manner of sickness and all manner of disease. Now the names of the twelve apostles are these who stood for the twelve tribes of Israel:

2. Peter, called Cephas, for the tribe of Reuben James, for the tribe of Naphtali; Thomas, called Dydimus, for the tribe of Zabulon; Matthew, called Levi for the tribe of Gad; John, for the tribe of Ephraim Simon, for the tribe of Issachar.

3. Andrew, for the tribe of Ioseph; Nathanael, for the tribe of Simeon; Thaddeus, for the tribe of Zabulon; Jacob, for the tribe of Benjamin; Jude, for the tribe of Dan; Philip, for the tribe of Asher. And Judas Iscariot, a Levite, who betrayed him, was also among them (but he was not of them). And Matthia and Barsabbas were also present with them.

4. Then he called in like manner twelve others to be Prophets, men of light to be with the Apostle and shew unto them the hidden things of God. And their names were Hermes, Aristobulus, Selenius, Nereus, Apollos, and Barsabbas; Andronicus, Lucius, Apelles, Zachaeus, Urbanus, and Clementos. And then he called twelve who should be Evangelists, and twelve who should be Pastors. A fourfold twelve did he call that he might send them forth to the twelve tribes of Israel, unto each, four.

5. And they stood around the Master, clad in white linen raiment, called to be a holy priesthood unto God for the service of the twelve tribes whereunto they should be sent.

6. These fourfold Twelve Iesus sent forth and charged them, saying, I will that ye be my Twelve Apostle with your companions, for a testimony into Israel. Go ye into the cities of Israel and to the lost sheep of the House of Israel. And as ye go, preach, saying, The kingdom of heaven is at hand. As I have baptized you in wader, so baptize ye them who believe.

7. Anoint and heal the sick, cleanse the lepers, raise the dead, cast out devils, freely ye have received, freely give. Provide neither gold, nor silver, nor brass in your purses. Nor scrip for your journey, neither two coats, neither shoes, nor yet staves; for the workman is worthy of his food; and eat that which is set before you, but of that which is gotten by taking of life, touch not, for it is not lawful to you.

8. And into whatsoever city or town ye shall enter, enquire who in it is worthy; and there abide till ye go thence. And when ye come into an house, salute it. And if the house be worthy, let your peace come upon it: but if it be not worthy, let your peace return to you.

9. Be ye wise as serpents and harmless as doves. Be ye innocent and undefiled. The Son of Man is: not come to destroy but to save, neither to take life, but to give life, to body and soul.

10. And fear not them which kill the body but are not able to kill the soul; but rather fear him who is able to destroy both soul and body in Gehenna.

11. Are not two sparrows sold for a farthing? and one of them shall not fall on the ground without permission of the All Holy. Yea, the very hairs of your head are all numbered. Fear yet not therefore, if God careth for the sparrow, shall he not care for you!

12. It is enough for disciples that they be as their master, and the servants as their lord. If they have called the master of the house Beelzebub, how much more shall they call them of his household? Fear them not therefore, for there is nothing covered, that shall not be revealed; or hid, that shall not be known.

13. What I tell you in darkness, that speak ye in light when the time cometh: and what ye hear in the ear, that preach ye upon the housetops. Whosoever therefore shall confess the truth before men, them will I confess also before my Parent Who is in heaven. But whosoever shall deny the truth before men, them will I also deny before my Parent Who is in heaven.

14. Verily I am come to send peace upon earth, but when I speak, behold a sword followeth. I am come to unite, but, behold, a man shall be at variance with his father, and the daughter with her mother, and the daughter-in-law with her mother-in-law. And a man's foes shall be they of his own household. For the unjust cannot mate with them that are just.

15. They who take not their cross and follow after me are not worthy of me. He that findeth his life shall lose it; and he that loseth his life for my sake, shall find it.

LECTION 17. 3. -Iudas Iscariot is here called a Levite. It may be symbolical of the fact, that the older priesthood was the bitter enemy of Iesus, the Prophets and the Priest of the newer Christian Dispensation.

v. 2-5. -Here we have a flood of light thrown on an obscure passage in Ephesians iv. 11, referring to an event of which there is no record whatever in the A. V. or in any other version of the Gospels which has come down to us. Plain enough is the passage in Ephesians as it stands, but obscure in its reference; and the only body of Christians who have in later times restored this ancient **fourfold ministry** is the "Catholic Apostolic Church," but with this difference, that what Iesus intended to be a permanent order, they have made only a lifetime institution, dying with the men that fill the office, at present the one left being removed by death. Under Iesus, the High Priest, or chief Shepherd and Bishop of the Universal Church, while the two and seventy afterward sent forth were the deacons in the higher ministry, altogether making the full number a hundred and twenty.

v. 6-9. -These words leave no doubt that the organization which Iesus first established was based on the older organization of the Yessenes (similar to that of the Buddhists), and from which have come the monasteries, friaries, and sisterhoods of the Christian Church, which have always been popular with the poor, and befriended them in times of trouble, and set them an example of Godly living; the corruptions and abuses, which set in now and then, being no argument against the use. They were a continual protest against the ways of the world, its vices and luxury and evil pursuits. "Leave all and follow me," was the continual call of the master, to those who could receive it.

Lection 18

The Sendeth Forth Of The Two and Seventy

1. AFTER these things the Lord appointed two and seventy also, and sent them two and two before his face into every city and place of the tribes whither he himself would come.

2. Therefore said he unto them, The harvest truly is great, but the labourers are few, pray ye therefore the Lord of the harvest that he would send forth labourers into the harvest.

3. Go your ways, behold I send you forth as lambs among wolves. Carry neither purse, nor scrip, nor shoes, and salute no man by the way.

4. And into whatsoever house ye enter, first say, Peace be to this house. And if the spirit of peace be there your peace shall rest upon it, if not it shall turn to you again.

5. And into whatsoever city ye enter, and they receive you, eat such things as are set before you without taking of life. And heal the sick that are therein, and say unto them, The kingdom of God is come nigh unto you.

6. And in the same house remain, eating and drinking such things as they give without shedding of blood, for the labourer is worthy of his hire. Go not from house to house.

7. But into whatsoever city ye enter and they receive you not, go your ways out into the streets of the same and say, Even the very dust of your city, which cleaveth on us, we do wipe off against you, notwithstanding be ye sure of this, that the kingdom of God is come nigh unto you.

8. Woe unto thee, Chorazin! woe unto thee, Bethsaida! for if the mighty works had been done in Tyre and Sidon, which have been done in you, they had a great while ago repented, sitting in sackcloth and ashes. But it shall be more tolerable for them in the judgement than for you.

9. And thou, Capernaum, which art exalted to heaven shalt be thrust down to hades. They that hear you, hear also me; and they that despise you, despise also me; and they that despise me, despise Him that sent me. But let all be persuaded in their own minds.

10. AND again Iesus said unto them: Be merciful, so shall ye obtain mercy. Forgive others, so shall ye be forgiven. With what measure ye mete, with the same shall it be meted unto you again.

11. As ye do unto others, so shall it be done you. As ye give, so shall it be given unto you. As ye judge others, so shall ye be judged. As ye serve others, so. shall ye be served.

12. For God is just, and rewardeth every one according to their works. That which they sow they shall also reap.

LECTION 18. 1 -This number (seventy-two), symbolizring amongst the Jews the Nations of the Earth, and late-denoting the Diaconate of the Church Universal (in priestly orders), was afterwards selected by the Christian Church as the complete number of its cardinals, as it had been before the number of members of the Jewish Sanhedrim.

Lection 19

Iesus Teacheth how to Pray
Error even in Prophets

1. As Iesus was praying in a certain place on a mountain, some of his disciples came unto him, and one of then said, Lord teach us how to pray. And Iesus said unto them, When thou prayest enter into thy secret chamber, and when thou hast closed the door, pray to Abba Amma Who is above and within thee, and thy Father-Mother Who seest all that is secret shall answer thee openly.

2. But when ye are gathered together, and pray in common, use not vain repetitions, for your heavenly Parent knoweth what things ye have need of before ye ask them. After this manner therefore pray ye:—

3. Our Father-Mother Who art above and within: Hallowed be Thy Name in twofold Trinity. In Wisdom, Love and Equity Thy Kingdom come to all. Thy will be done, As in Heaven so in Earth. Give us day by day to partake of Thy holy Bread, and the fruit of the living Vine. As Thou dost forgive us our trespasses, so may we forgive others who trespass against us. Shew upon us Thy goodness, that to others we may shew the same. In the hour of temptation, deliver us from evil.

4. Shew upon us Thy goodness, that to others we may shew the same. In the hour of temptation, deliver us from evil.

5. And wheresoever there are seven gathered together in My Name there am I in the midst of them; yea, if only there be three or two; and where there is but one who prayeth in secret, I am with that one.

The Gospel Of The Holy 12

6. Raise the Stone, and there thou shall find me. Cleave the wood, and there am I. For in the fire and in the water even as in every living form, God is manifest as it's Life and it's Substance.

7. AND the Lord said, If thy brother hath sinned in word seven times a day, and seven times a day hath made amendment, receive him. Simon said to him, Seven times a day?

8. The Lord answered and said to him, I tell thee also unto seventy times seven, for even in the Prophets, after they were anointed by the Spirits utterance of sin was found.

9. Be ye therefore considerate, be tender, be ye pitiful, be ye kind, not to your own kind alone, but to every creature which is within your care, for ye are to them as gods, to whom they look in their need. Be ye slow to anger for many sin in anger which they repented of, when their anger was past.

10. AND there was a man whose hand was withered and he came to Iesus and said, Lord, I was a mason seeking sustenance by my hands, I beseech thee restore to me my health that I may not beg for food with shame. And Iesus healed him, saying There is a house made without hands, seek that thou mayest dwell therein.

LECTION 19. 2 -There are two versions given of the Lord's Prayer, this one, the fullest, being given to the Twelve and their companions, and a shorter form afterwards to the people in his Sermon on the Mount.

v. 5, 6 -An ancient saying, long lost to the Church. The all-pervading nature of Deity seems plainly taught, which, in a recently recovered fragment, is obscure.

Lection 20

The Return of the Two and Seventy

1. AND after a season the two and seventy returned again with joy, saying, Lord, even the demons are subject unto us through thy name.

2. And he said unto them, I beheld Satan as lightning fall from heaven.

3. Behold I give unto you power to tread on serpents and scorpions, and over all the power of the enemy; and nothing shall by any means hurt you. Notwithstanding in this, rejoice not, that the spirits are subject unto you; but rather rejoice, because your names are written in Heaven.

4. In that hour Iesus rejoiced in spirit, and said I thank thee, Holy Parent of heaven and earth, that thou hast hid these things from the wise and prudent, and hast revealed them unto babes: even so, All Holy, for so it seemed good in thy sight.

5. All things are delivered to me of the All-Parent: and no man knoweth the Son who is the Daughter, but the All Parent; nor who the All-Parent is, but the Son even the Daughter, and they to whom the Son and the Daughter will reveal it.

6. And he turned him unto his disciples, and said privately, Blessed are the eyes which see the things that ye see. For I tell you, that many prophets and kings have desired to see those things which ye see, and have not seen them; and to hear those things which ye hear, and have not heard them.

7. Blessed are ye of the inner circle who hear my word and to whom mysteries are revealed, who give to no innocent creature the pain of prison or of death, but seek the good of all, for to such is everlasting life.

8. Blessed are ye who abstain from all things gotten by bloodshed and death, and fulfill all righteousness: Blessed are ye, for ye shall attain to Beatitude.

LECTION 20. 9 -There is no trace of the events between Lections 18 and 20 other than this giving of the form of Prayer as a model for all time.

The Gospel of the Holy Twelve

Translated from the original Aramaic
by Rev. G.J.R. Ouseley

Section 3, Lections 21 thru 30

Lection 21

Iesus Rebuketh Cruelty to a Horse.
Condemneth the Service of Mammon.

1. AND it came to pass that the Lord departed from the City and went over the mountains with this disciples. And they came to a mountain whose ways were steep and there they found a man with a beast of burden.

2. But the horse had fallen down, for it was over laden, and he struck it till the blood flowed. And Jesus went to him and said: "Son of cruelty, why strikest thou thy beast? Seest thou not that it is too weak for its burden, and knowest thou not that it suffereth?"

3. But the man answered and said: "What hast thou to do therewith? I may strike it as much as it pleaseth me, for it is mine own, and I bought it with a goodly sum of money. Ask them who are with thee, for they are of mine acquaintance and know thereof."

4. And some of the disciples answered and said: Yea, Lord, it is as he saith, We have seen when he bought it. And the Lord said again "See ye not then how it bleedeth, and hear ye not also how it waileth and lamenteth ?" But they answered and said: "Nay, Lord, we hear not that it waileth and lamenteth? "

5. And the Lord was sorrowful, and said: "Woe unto you because of the dullness of your hearts, ye hear not how it lamenteth and crieth unto the heavenly Creator for mercy, but thrice woe unto him against whom it crieth and waileth in its pain."

6. And he went forward and touched it, and the horse stood up, and its wounds were healed. But to the man he said: "Go now thy way and strike it henceforth no more, if thou also desireth to find mercy."

7. AND seeing the people come unto him, Jesus, said unto his disciples, Because of the sick I am sick; because of the hungry I am hungry; because of the thirsty I am athirst.

8. He also said, I am come to end the sacrifices and feasts of blood, and if ye cease not offering and eating of flesh and blood, the wrath of God shall not cease from you, even as it came to your fathers in the wilderness, who lusted for flesh, and they eat to their content, and were filled with rottenness, and the plague consumed them.

9. And I say unto you, Though ye be gathered together in my bosom, if ye keep not my commandments I will cast you forth. For if ye keep not the lesser mysteries, who shall give you the greater.

10. He that is faithful in that which is least is faithful also in much: and he that is unjust in the least is unjust also in much.

11. If therefore ye have not been faithful in the mammon of unrighteousness, who will commit to your trust the true riches? And if ye have not been faithful in that which is another man's, who shall give you that which is your own ?

12. No servant can serve two masters: for either he will hate the one, and love the other; or else he will hold to the one and despise the other. **Ye cannot serve God and mammon.** And the Pharisees also, who were covetous, heard all these things, and they derided him.

13. And he said unto them, Ye are they which justify yourselves before men; but God knoweth your hearts: **for that which is highly esteemed among men is abomination in the sight of God.**

14. The law and the prophets were until John; since that time the kingdom of God is preached, and every man presseth into it. But it is easier for heaven and earth to pass away, than one title of the law to fail.

15. Then there came some women to him and brought their infants unto him, to whom they yet gave suck at their breasts, that he should bless them; and some said, Why trouble ye the master?

16. But Iesus rebuked them, and said, Of such will come forth those who shall yet confess me before men. And he took them up in his arms and blessed them.

LECTION 21. 2-6. -This touching incident is to be found also in a very ancient Coptic fragment of the Life of Iesus- others of a like nature also recorded in their places in this Gospel, show how he, the Divine Saviour of the world, regarded the ill-treatment of the "lower" animals as a **grievous sin.**

v. 12. -The divine love of Iesus for all God's creatures is everywhere evidenced by this Gospel, and his belief that all life is one, is abundantly justified by the teaching of true modern science, physical and occult.

Lection 22

The Restoration Of Iairus' Daughter

1. AND behold there cometh one of the rulers of the synagogue, Iairus by name; and when he saw him, he fell at his feet, and he besought him greatly, saying, My little daughter lieth at the point of death; I pray thee, come and lay thy hands on her, that she may be healed, and she shall live. And Jesus went with him, and much people followed him and thronged him.

2. AND a certain woman, which had an issue of blood twelve years, and had suffered many things of many physicians, and had spent all that she had, and was nothing bettered, but rather grew worse.

3. When she had heard of Iesus, she came in the press behind and touched his garments For she said, If I may touch but his garment, I shall be whole. find straightway the fountain of her blood was dried up; and she felt in her body that she was healed of that plague.

4. And Iesus, immediately knowing in himself that virtue had gone out of him, turned him about in the press and said, Who touched my vesture? And his disciples said unto him, Thou seest the multitude thronging thee and sayeth thou, Who touched me?

5. And he looked round about to see her that had done this thing. But the woman, fearing and trembling, knowing what was done in her, came and fell down before him and told him all the truth. And he said unto her, Daughter, thy faith hath made thee whole; go in peace and be whole of thy plague.

6. WHILE he yet spake, there came from the ruler of the synagogue's house certain which said, Thy daughter is dead: why troublest thou the Master any further?

7. As soon as Iesus heard the word that was spoken, he saith unto the ruler of the synagogue, Be not afraid, only believe. And he suffered no man to follow him save Peter and James and John the brother of James.

8. And he cometh to the house of the ruler of the synagogue, and seeth the tumult and the minstrels, and them that lamented and wailed greatly.

9. And when he was come in he said unto him, Why make ye this ado and weep? The damsel is not dead but sleepeth. And they laughed him to scorn, for they thought she was dead, and believed him not. But when he had put them all out, he taketh two of his disciples with him, and entered in where the damsel was lying.

10. And he took the damsel by the hand and said unto her, Talitha cumi; which is, being interpreted, Damsel, I say unto thee arise.

11. And straightway the damsel arose and walked. And she was of the age of twelve years. And they were astonished with a great astonishment.

12. And he charged them straightly that no man should make it known, and commanded that something should be given to her to eat.

LECTION 22 -The daily increasing discoveries in modern times of cases of trance or of suspended animation, in which those carried to burial certified as dead by medical men have revived, suggest the thought how much more numerous must have been such cases in days when medical science knew little or nothing of the symptoms of real death. When it is now ascertained that five per thousand on an average are restored to life who have been certified dead or carried to burial, how many more such cases must have occurred in those times when true physicians and magnetic healers were looked upon almost as gods?

Lection 23

Iesus And The Samaritan Woman

1. THEN cometh Iesus to a city of Samaria, which is called Sychar, near to the parcel of ground that Jacob gave to his son Joseph.

2. Now Jacob's well was there. Iesus therefore, being wearied with his journey, sat alone on the edge of the well, and it was about the sixth hour.

3. And there cometh a woman of Samaria to draw water; Iesus saith unto her, Give me to drink. (For his disciples were gone away unto the city to buy food).

4. Then saith the woman of Samaria unto him, How is it that thou being a Jew, asketh drink of me, who am a woman of Samaria? (for the Jews have no dealings with the Samaritans.)

5. Iesus answered and said unto her, If thou knewest the gift of God and who it is that saith to thee, Give me drink, thou wouldest have asked of God, who would have given thee living water.

6. The woman saith unto him, Sir, thou hast nothing to draw with, and the well is deep, from whence hast thou that living water. Art thou greater than our father Jacob, who gave us the well and drank thereof, himself and his children and his camels and oxen and sheep.

7. Iesus answered and said unto her, Whosoever drinketh of this water shall thirst again, but whosoever drinketh of the water that I shall give him shall never thirst; but the water that I shall give him shall be in him a well of water springing up into everlasting life.

8. The woman saith unto him, Sir, give me this water, that I thirst not, neither come hither to draw. Iesus saith unto her, Go, call thy husband and come hither. The woman answered and said, I have no husband.

9. Iesus looking upon her, answered and said unto her, Thou hast well said, I have no husband. For thou hast had five husbands and he whom thou now hast is not called thy husband, in that saidst thou truly.

10. The woman saith unto him, Sir, I perceive that thou art a prophet. Our fathers worshipped in this mountain and ye say that

in Jerusalem is the place where men ought to worship.

11. Iesus saith unto her, Woman, believe me, the hour cometh, when ye shall neither in this mountain nor yet at Jerusalem worship God. Ye worship ye know not what; we know what we worship; for salvation is of Israel.

12. But the hour cometh and now is, when the true worshippers shall worship the All-Parent in spirit and in truth; for such worshippers the All-Holy seeketh. God is a Spirit and they that worship, must worship in spirit and in truth.

13. The woman saith unto him, I know that Messiah cometh who is called the Christ: when he is come he will tell us all things. Jesus saith unto her, I am he Who speaketh unto thee.

14. And upon this came his disciples and marveled that he talked with the woman, yet no man said, What seekest thou ? or, Why talkest thou with her?

15. The woman then left her waterpot, and went her way into the city and saith unto the men, Come, see a man which told me all things that ever I did: is not this the Christ?

16. Then they went out of the city and came unto him, and many of the Samaritans believed on him, and they besought him that he would tarry with them; and he abode there two days.

LECTION 23. 1-13. -A similar event is recorded in the life of Buddha, where he asks water of a woman, and receives it from a woman of lower caste, who asks how he, of a higher caste, a Brahmin, comes to ask water of one so much lower. It should cast no doubt on this passage.

Lection 24

Iesus Denounces Cruelty
He Healeth the Sick

1. As Iesus passed through a certain village he saw a crowd of idlers of the baser sort, and they were tormenting a cat which they had found and shamefully treating it. And Iesus commanded them to desist and began to reason with them, but they would have none of his words, and reviled him.

2. Then he made a whip of knotted cords and drove them away, saying, This earth which my Father-Mother made for joy and gladness, ye have made into the lowest hell with your deeds of violence and cruelty; And they fled before his face.

3. But one more vile than the rest returned and defied him. And Iesus put forth his hand, and the young man's arm weathered, and great fear came upon all; and one said, He is a sorcerer.

4. And the next day the mother of the young man came unto Iesus, praying that he would restore the withered arm. And Iesus spake unto them of the law of love and the unity of all life in the one family of God. And he also said, As ye do in this life to your fellow creatures, so shall it be done to you in the life to come.

5. And the young man believed and confessed his sins, and Iesus stretched forth his hand, and his withered arm became whole even as the other, And the people glorified God who had given such power unto man.

6. AND when Iesus departed thence, two blind men followed him, crying and saying, Thou son of David, have mercy on us. And when he was come into the house the blind men came to him, and Iesus saith unto them, Believe ye that I am able to do this?

7. They said unto him, Yea, Lord. Then touched he their eyes, saying, According to your faith be it unto you. And their eyes were opened, and Iesus straitly charged them, saying, See that ye tell no man, But they, when they were departed, spread abroad his fame in all that country.

8. As they went forth, behold, they brought to him a dumb man possessed with a demon. And when the demon was cast out the dumb spake, and the multitude marvelled, saying, It was never so seen in Israel. But the Pharisees said, He casteth out demons through the prince of the demons.

9. AND Iesus went about all the cities and villages, teaching in their synagogues and preaching the gospel of the kingdom and healing every sickness and every disease among the people.

10. But when he saw the multitudes he was moved with compassion on them, because they fainted and were scattered abroad, as sheep having no shepherd.

11. Then said he unto his disciples, The harvest truly is plentiful, but the labourers are few; pray ye therefore the Lord of the harvest, that he will send forth labourers into his harvest.

12. AND his disciples brought him two small baskets with bread and fruit, and a pitcher of water. And Iesus set the bread and the fruit before them and also the water. And they did eat and drink and were filled.

13. And they marvelled, for each had enough and to spare, and there were four thousand. And they departed blessing God for what they had heard and seen.

LECTION 24. 1-5. -The cat was an ancient symbol of Deity, on account of its seeing in the dark and otter attributes. More than one instance is given of Iesus' protection of these beautiful animals which in Iudea were, as they are even now in some places, unjustly despised and regarded with disfavour. He, the Friend of all things that suffered, cast his protection round these innocent creatures, teaching men and women to do likewise, and to feel for all the weak and oppressed. This beautiful and much maligned animal was a native of Egypt. But there is no difficulty here, for Egyptian families visited Palestine, and would naturally bring their venerated animals with them, not leave them to neglect or worse, as some "Christians" who ought to know better.

Lection 25

The Sermon On The Mount (Part I)

1. Iesus seeing the multitudes, went up into a mountain: and when he was seated, the twelve came unto him, and he lifted up his eyes on his disciples and said:

2. Blessed in spirit are the poor, for theirs is the kingdom of heaven. Blessed are they that mourn: for they shall be comforted. Blessed are the meek; for they shall inherit the earth. Blessed are they who do hunger and thirst after righteousness: for they shall be filled.

3. Blessed are the merciful: for they shall obtain mercy. Blessed are the pure in heart: for they shall see God. Blessed are the peacemakers: for they shall be called the children of God. Blessed are they which are persecuted for righteousness sake: for theirs is the kingdom of God.

4. Yea, blessed are ye, when men shall hate you' and when they shall separate you from their company, and shall reproach you, and cast out your name as evil, for the Son of man's sake. Rejoice ye in that day, and leap for joy: for, behold, your reward is great in heaven; for in the like manner did their fathers unto the prophets.

5. Woe unto you that are rich! for ye have received in this life your consolation. Woe unto you that are full! for ye shall hunger. Woe unto you that laugh now! for ye shall mourn and weep. Woe unto you when all men shall speak well of you' for so did their fathers to the false prophets.

6. Ye are the salt of the earth, for every sacrifice must be salted with salt, but if the salt have lost its savour, wherewith shall it be salted? it is thenceforth good for nothing, but to be cast out, and to be trodden under foot.

7. Ye are the light of the world. A city that is built on a hill cannot be hid. Neither do men light a candle, and put it under a bushel, but on a candlestick; and it giveth light unto all that are in the house. Let your light so shine before men, that they may see your good works, and glorify your Parent who is in heaven.

8. Think not that I am come to destroy the law, or the prophets: I am not come to destroy, but to fulfill. For verily I say unto you, Till heaven and earth pass, one jot or one tittle shall in no way pass from the law or the prophets till all be fulfilled. But behold One greater than Moses is here. and he will give you the higher law, even the perfect Law, and this Law shall ye obey.

9. Whosoever therefore shall break one of these commandments which he shall give, and shall teach men so, they shall be called the least in the kingdom; but whosoever shall do, and teach them, the same shall be called great in the kingdom of Heaven.

10. Verily they who believe and obey shall save their souls, and they who obey not shall lose them. For I say unto you, That except your righteousness shall, exceed the righteousness of the scribes and Pharisees ye shall not enter the kingdom of Heaven.

11. Therefore if thou bring thy gift to the altar and there rememberest that thy brother hath aught against thee, leave there thy gift before the altar, and go thy way; first be reconciled to thy brother, and then come and offer thy gift.

12. Agree with thine adversary quickly, while thou art in the way with him; lest at any time thy adversary deliver thee to the Judge, and the judge deliver thee to the officer, and thou be cast into prison. Verily I say unto thee. Thou shalt by no means come out thence till thou hast paid the uttermost farthing.

13. Ye have heard that it hath been said, Thou shalt love thy neighbour and hate thine enemy. But I say unto you which hear, Love your enemies, do good to them which hate you.

14. Bless them that curse you, and pray for them which despitefully use you. That ye may be the children of your Parent Who maketh the sun to rise on the evil and the good, and sendeth rain on the Just and on the unjust.

15. For if ye love them which love you what thank have ye? for sinners also love those that love them. And if ye do good to them which do good to you, what thank have ye? for sinners even do the same. And if ye salute your brethren only, what do ye more than others? do not even so the taxgatherers?

16. And if a desire be unto thee as thy life, and it turn thee from the truth, cast it out from thee, for it is better to enter life possessing truth, than losing it, to be cast into outer darkness.

17. And if that seem desirable to thee which costs another pain or sorrow, cast it out of thine heart; so shalt thou attain to peace. Better it is to endure sorrow, than to inflict it, on those who are weaker.

18. Be ye therefore perfect, even as your Parent Who is in heaven is perfect.

LECTION 25. 2. -It is remarkable how persistent has been the false rendering of these words in the received Gospels. It is too evident to need any comment. It is not poverty of spirit that Christ commended, but the spiritual effects of literal poverty (not pauperism), which are more frequent than those of abundant riches.

v. 6-7. -Suggestive is this passage of the custom of the Christian Church in building their monasteries and convents generally on high places, the bands of holy men and women therein being truly, in the Dark Ages, the salt of the earth, the light on a hill, without which society would have rotted to the core, and been universally corrupt. The occasional abuses argue nothing against their more blessed influences. Without them our Scriptures would not have been preserved, even in their present condition, and civilization would have been extinct. To the monks of S. Basil and S. Benedict are due the remains of Christianity that have been handed down to us, and by such institutions rationally conducted will Christianity be revived in a higher and purer form, and the Scriptures restored to their original purity, as well as the ancient worship of God. The laxity of some modern monasteries is to be regretted in the matter of flesh-eating, under the plea of health, there being really no such necessity with the abundance of food from the vegetable world as well as animal products. The Carthusian and other monasteries stand as a noble testimony to the healthfulness of the rule when observed in strictness and unabated rigour.

Lection 26

The Sermon On The Mount (Part II)

1. TAKE heed that ye do not your alms before men, to be seen of them: otherwise ye have no reward of your Parent who is in heaven. Therefore when thou doest thine alms, do not sound a trumpet before thee, as the hypocrites do in the synagogues and in the streets, that they may have glory of men. Verily I say unto you, they have their reward.

2. But when thou givest alms, let not thy left hand know what thy right hand doeth, and take heed that thine alms may be in secret; and the Secret One which seest in secret shall approve then openly.

3. And when thou prayest, thou shalt not be as the hypocrites are: for they love to pray standing in the synagogues and on the corners of the streets that they may be seen of men. Verily I say unto you, They have their reward.

4. But thou, when thou prayest enter into thy chamber and when thou hast shut thy door pray to thy Father-Mother who is in secret; and the secret One that seeth in secret shall approve thee openly.

5. And when ye pray in common, use not vain petitions, as the heathen do: for they think that they shall be heard for their much speaking. Be not ye therefore like unto them: for your heavenly Parent knoweth what things ye have need of, before ye ask After this manner therefore pray ye, when ye are gathered together:

6. Our Parent Who art in heaven: Hallowed be Thy Name. Thy kingdom come. Thy will be done; in earth as it is in heaven. Give us day by day our daily bread, and the fruit of the living Vine. As Thou forgivest us our trespasses, so may we forgive the trespasses of others. Leave us not in temptation. Deliver us from evil. Amun.

7. For if ye forgive men their trespasses, your heavenly Parent will also forgive you: but if ye forgive not men their trespasses, neither will your Parent in heaven forgive you your trespasses.

8. Moreover when ye fast, be not, as the hypocrites, of a sad countenance; for they disfigure their faces, that they may appear unto men to fast. Verily I say unto you, they have their reward.

9. And I say unto you, Except ye fast from the world and its evil ways, ye shall in no wise find the Kingdom; and except ye keep the Sabbath and cease your haste to gather riches, ye shall not see the Father-Mother in heaven. But thou, when thou fastest, anoint thine head and wash thy face, that thou appear not unto men to fast, and the Holy One who seeth in secret will approve thee openly.

10. Likewise also do ye, when ye mourn for the dead and are sad, for your loss in their gain. Be not as those who mourn before men and make loud lamentation and rend their garments, that they may be seen of men to mourn. For all souls are in the hands of God, and they who have done good, do rest with your ancestors in the bosom of the Eternal.

11. Pray ye rather for their rest and advancement, and consider that they are in the land of rest, which the Eternal hath prepared for them, and have the just reward of their deeds, and murmur not as those without hope.

12. Lay not up for yourselves treasures upon earth, where moth and rust doth corrupt, and where thieves break through and steal; but lay up for yourselves treasures in heaven, where neither moth not rust doth corrupt and where thieves do not break through nor steal. For where your treasure is, there will your heart be also.

13. The lamps of the body are the eyes: if therefore thy sight be clear, thy whole body shall be full of light. But if thine eyes be dim or lacking, thy whole body shall be full of darkness. If therefore the light that is in thee be darkness, how great is that darkness!

14. No man can serve two masters; for either he will hate the one and love the other; or else he will hold to the one and despise the other. Ye cannot serve God and mammon.

15. Therefore I say unto you, Be not over anxious for your life what ye shall eat, or what ye shall drink; nor yet for your body, what ye shall put on. Is not the life more than meat and the body than raiment? And what shall it profit a man if he gain the whole world and lose his life ?

16. Behold the fowls of the air; for they sow not, neither do they reap, nor gather into barns; yet your heavenly Parent feedeth them. Are ye not much better cared for than they? Which of you by taking thought can add one cubit unto his stature? And why spend all your thought for raiment ? Consider the lilies of the field, how they grow; they toil not, neither do they spin. And yet I say unto you, Solomon in all his glory was not arrayed like one of these.

17. Wherefore shall not God who clothes the grass of the field, which to day is, and tomorrow is cast into the oven, much more clothe you, O ye of little faith?

18. Therefore be not over anxious, saying, What shall we eat? or, What shall we drink? or, Wherewithal shall we be clothed? (all Which things do the Gentiles seek). For your heavenly Parent knoweth that ye have need of all these things. But seek ye first the kingdom of God and its righteousness and all these things shall be added unto you. Meet not in advance the evils of the morrow; sufficient unto the day is the evil thereof.

LECTION 26. 9. -Meaning; that if the vision be set on one single object and no other, great is the clearness of vision; while, if the eyes be set on number of other objects, the clearness will be diminished with regard to that one.

Lection 27

The Sermon On The Mount (Part III)

1. JUDGE not, that ye be not judged. For with what judgment ye judge, ye shall be judged: and with what measure ye mete, it shall be measured to you again; and as ye do unto others, so shall it be done unto you.

2. And why beholdest thou the mote that is in thy brother's eye, but considerest not the beam that is in thine own eye? Or how wilt thou say to thy brother, Let me pull the mote out of thine eye; and behold a beam is in thine own eye? Thou hypocrite, first cast the beam out of thine own eye; and then shall thou see clearly to cast the mote out of thy brother's eye.

3. Give not that which is holy unto the dogs' neither cast ye your pearls before swine; lest they trample them under their feet and turn again and rend you.

4. Ask and it shall be given you; seek, and ye shall find; knock, and it shall be opened unto you: for everyone that asketh receiveth, and he that seeketh findeth, and to them that knock it shall be opened.

5. What man is there of you who, if his child ask bread, will give it a stone? Or, if it ask a fish, will give it a serpent? If ye then, being evil, know how to give good gifts unto your children, how much more shall your Parent Who is in heaven give good things to them that ask?

6. Therefore all things whatsoever ye would that men should do to you, do ye even so to them. And what ye would not that men should do unto you, do ye not so unto them; for this is the Law and the prophets.

7. Enter ye in at the strait gate, for strait is the way and narrow the gate that leadeth unto life, and few there be that find it. But wide is the gate and broad is the way that leadeth to destruction, and many there be who go in thereat.

8. Beware of false prophets, which come to you in sheep's clothing, but inwardly are ravening wolves. Ye shall know them by their fruits. Do men gather grapes of thorns, or figs of thistles?

9. Even so, every good tree bringeth forth good fruit, but a corrupt tree bringeth forth evil fruit. Every tree that bringeth not forth good fruit is only fit to be hewn down and cast into the fire. Wherefore by their fruits ye shall know the good from the evil.

10. Not every one that saith unto me, Lord, Lord, shall enter into the kingdom of heaven; but he that doeth the will of my Father-Mother Who is in heaven. Many will say to me in that day, Lord, Lord, have we not prophesied in thy Name? and in thy Name have cast out devils? and in thy Name done many wonderful works? And then will I say unto them, I never knew you: depart from me, ye that work iniquity.

11. Therefore whosoever heareth these sayings of mine, and doeth them, I will liken him unto a wise man who built his house foursquare upon a rock. And the rain descended, and the floods came, and the winds blew upon that house; and it fell not, for it was founded upon a rock.

12. And everyone that heareth these sayings of mine, and doeth them not, shall be likened unto a foolish man, who built his house upon the sand, and the rain descended, and the floods came and the winds blew and beat upon that house, and it fell, and great was the fall of it. But a city which is built foursquare, enclosed in a circle or on the top of a hill, and established on a rock, can neither fall nor be hidden.

13. And it came to pass, when Iesus had ended these sayings, the people were astonished at his doctrine. For he taught them as one appealing to the reason and the heart, and not as the scribes who taught rather by authority.

LECTION 27. 2. -The sin of hypocrisy is most loathsome, and the most difficult for those to see who are vitiated by it. To condemn in others the sins we practice ourselves is a common sin of society, which Christ ever reprobates.

LECTION 27. 12. -Note the importance which these symbols (including the equilateral triangle) possessed in the eyes of Iesus as illustrations of Eternal truths. The slight mention of these shows the Gospel was written, or addressed to people well acquainted with the mysteries they represent.

Lection 28

Iesus Releases The Rabbits And Pigeons

1. IT came to pass one day as Iesus had finished his discourse, in a place near Tiberias where there are seven wells, a certain young man brought live rabbits and pigeons, that he might have to eat with his disciples.

2. And Iesus looked on the young man with love and said to him, Thou hast a good heart and God shall give thee light, but knowest thou not that God in the beginning gave to man the fruits of the earth for food, and did not make him lower than the ox, or the horse, or the sheep, that he should kill and eat the flesh and blood of his fellow creatures.

3. Ye believe that Moses indeed commanded such creatures to be slain and offered in sacrifice and eaten, and so do ye in the Temple, but behold a greater than Moses is herein and he cometh to put away the bloody sacrifices of the law, and the feasts on them, and to restore to you the pure oblation and unbloody sacrifice as in the beginning, even the grains and fruits of the earth.

4. Of that which ye offer undo God in purity shall ye eat, but of that kind which ye offer not in purity shall ye not eat, for the hour cometh when your sacrifices and feasts of blood shall cease, and ye shall worship God with a holy worship and a pure Oblation.

5. Let these creatures therefore go free, that they may rejoice in God and bring no guilt to man. And the young man set them free, and Iesus break their cages and their bonds.

6. But lo, they feared lest they should again be taken captive, and they went not away from him, but he spake unto them and dismissed them, and they obeyed his word, and departed in gladness.

7. AT that time as they sat by the well, which was in the midst of the six Iesus stood up and cried out, If any are thirsty, let them come unto me and drink, for I will give to them of the waters of life.

8. They who believe in me, out of their hearts shall flow rivers of water, and that which is given unto them shall they speak with power, and their doctrine shall be as living water.

9. (This he spake of the Spirit, which they that believed on him should receive, for the fullness of the Spirit was not yet given because that Iesus was not yet glorified).

10. Whosoever drinketh of the water that I shall give shall never thirst, but the water which cometh from God shall be in them a well of water, springing up unto everlasting life.

11. AND at that time John sent two of his disciples, saying, Art thou he that should come, or look we for another? and in that same hour he cured many of their infirmities and plagues, and of evil spirits, and unto many blind, he gave sight.

12. Then Iesus answering said unto them, Go your way, and tell John what things ye have seen and heard; how that the blind see, the lame walk, the lepers are cleansed, the deaf hear, the dead are raised, to the poor the gospel is preached. And blessed is he, whosoever shall not be offended in me.

13. And when the messengers of John were departed, he began to speak unto the people concerning John, What went ye out into the wilderness for to see? A reed shaken with the wind, or a man clothed in soft raiment? Behold, they which are georgeously apparelled, and live delicately, are in kings' courts.

14. But what went ye out for to see? A prophet Yea, I say unto you, and the greatest of prophets.

15. This is he, of whom it is written, Behold, I send my messenger before thy face, which shall prepare thy way before thee. For I say unto you, Among those that are born of women, there is not a greater prophet than John the Baptist.

16. And all the people that heard him, and the taxgatherers, justified God, being baptized with the baptism of John. But the Pharisees and lawyers rejected the counsel of God against themselves, being not baptized of him.

LECTION 28. 1-5. -It is easy to see how this would have horrified the mind of Iesus had he lived in these semi- heathen times.

LECTION 28. 16. -To exalt unduly the Christian Sacrament of Baptism, certain words have been interpolated in the A.V. The context shows that such was not the intended meaning, for immediately after, they "justified God by being baptized with Iohn's Baptism."

Lection 29

The feeding of the Five Thousand With Six Loaves and Seven Clusters Of Grapes.
Healing Of The Sick

1. AND the Feast of the Passover drew nigh, and the Apostles and their fellows gathered themselves together unto Iesus and told him all things, both what they had done and what they had taught. And he said unto them, Come ye yourselves apart into a desert place and rest a while: for there were many coming and going, and they had no leisure so much as to eat.

2. And they departed into a desert place by ship privately. And the people saw them departing, and many knew him, and ran afoot thither out of all cities, and outwent them, and came together unto him.

3. And Iesus, when he came forth, saw much people and was moved with compassion towards them, because they were as sheep having not a shepherd.

4. And the day was far spent, and his disciples came unto him and said, This is a desert place, and now the time is far passed. Send them away, that they may go into the country round about into the villages, and buy themselves bread, for they have nothing to eat.

5. He answered and said unto them, Give ye them to eat. And they say unto him, Shall we go and buy two hundred pennyworth of bread, and give them to eat ?

6. He saith unto them, How many loaves have ye? go and see. And when they knew, they said, Six loaves and seven clusters of grapes. And he commanded them to make all sit down by companies of fifty upon the grass. And they sat down in ranks by hundreds and by fifties.

7. And when he had taken the six loaves and the seven clusters of grapes, he looked up to heaven, and blessed and brake the loaves, and the grapes also and gave them to his disciples to set before them and they divided them among them all.

8. And they did all eat and were filled. And they took up twelve baskets full of the fragments that were left. And they that did eat of the loaves and of the fruit were about five thousand men, women and children, and he taught them many things.

9. And when the people had seen and heard, they were filled with gladness and said, Truly this is that Prophet that should come into the world. And when he perceived that they would take him by force to make him a king, he straightway constrained his disciples to get into the ship, and to go to the other side before him unto Bethsaida, while he sent away the people.

10. And when he had sent them away he departed into a mountain to pray. And when even was come, he was there alone, but the ship was now in the midst of the sea, tossed with waves, for the wind was contrary.

11. The third watch of the night Iesus went unto them, walking on the sea. And when the disciples saw him walking on the sea, they were troubled, saying, It is a spirit; and they cried out for fear. But straightway Iesus spake unto them, saying. Be of good cheer; it is I; be not afraid.

12. And Peter answered him and said, Lord, if it be thou, bid me come unto thee on the water. And he said, Come. And when Peter was come down out of the ship, he walked on the water, to go to Iesus. But when he saw the wind boisterous, he was afraid, and beginning to sink, he cried, saying, Lord, save me.

13. And immediately Iesus stretched forth his hand, and caught him, and said unto him, O thou of little faith, wherefore didst thou doubt? For did I not call thee ?

14. And he went up unto them into the ship, and the wind ceased, and they were sore amazed in themselves beyond measure and wondered. For they considered not the miracle of the loaves and the fruit, for their heart was hardened.

15. And when they were come into the ship there was a great calm. Then they that were in the ship came and worshipped him, saying, Of a truth thou art a Son of God.

16. And when they had passed over, they came unto the land of Gennesaret and drew to the shore And when they were come out of the ship straightway they knew him. And ran through that whole region round about, and began to carry about in beds, those that were sick, where they heard he was.

17. And withersoever he entered, into villages, or cities, or country, they laid the sick in the streets, and besought him that they might touch if it were but the border of his garment, and as many as touched him were made whole.

18. After these things Iesus came with his disciples into Judea, and there he tarried and baptized many who came unto him and received his doctrine.

LECTION 29. -The feeding of five thousand with five loaves and seven clusters of grapes has a deep mystical significance, which space forbids to enter on here, but the wise will understand. The two numbers, *e.g*, symbolize Matter and Spirit, Bread and Wine, Substance and Life.

Lection 30

The Bread Of Life And The Living Vine

1. THE day following, the people which stood on the other side of the sea, saw that there had been no other boat there, save the one whereinto his disciples had entered and that Iesus went not with his disciples into the boat, but that his disciples were gone alone. And when the people therefore saw that Iesus was not there, neither his disciples, they also took ship and came to Capernaum, seeking for Iesus.

2. And when they had found him on the other side of the sea, they said unto him, Rabbi, how camest thou hither? Yeshua answered them and said, Verily, verily, I say unto you, ye seek me, not because ye saw the miracles, but because ye did eat of the loaves and the fruit, and were filled. Labour not for the meat which perisheth, but for that meat which endureth unto everlasting life, which the Son of Man, Who is also the Child of God, shall give unto you, for him hath God the All Parent sealed.

3. Then said they unto him, What shall we do that we may work the works of God? Iesus answered and said unto them, This is the work of God, that ye believe the truth, in me who am, and who giveth unto you, the Truth and the Life.

4. They said therefore unto him, What sign shewest thou then that we may see and believe thee? What dost thou work? Our fathers did eat manna in the desert; as it is written, He gave them bread from heaven to eat.

5. Then Iesus said unto them, Verily, verily, I say unto you, Moses gave you not the true bread from heaven, but my Parent giveth you the true bread from heaven and the fruit of the living vine. For the food of God is that which cometh down from heaven, and giveth life unto the world.

6. Then said they unto him, Lord, evermore give us this bread, and this fruit. And Iesus said unto them, I am the true Bread, I am the living Vine, they that come to me shall never hunger; and they that believe on me shall never thirst. And verily I say unto you, Except ye eat the flesh and drink the blood of God, ye have no life in you. But ye have seen me and believe not.

7. All that my Parent hath given to me shall come to me and they that come to me I will in no wise cast out. For I came down from heaven, not to do mine own will, but the will of God who sent me. And this is the will of God who hath sent me, that of all which are given unto me I should lose none, but should raise them up again at the last day.

8. The Jews then murmured at him, because he said I am the bread which cometh down from heaven. And they said, Is not this Iesus, the son of Joseph and Mary whose parentage we know? how is it then that he saith, I came down from heaven?

9. Iesus therefore answered and said unto them, Murmur ye not among yourselves. None can come to me except holy Love and Wisdom draw them, and these shall rise at the last day. It is written in the prophets, They shall be all taught of God. Every man therefore that hath heard and hath learned of the Truth, cometh unto me.

10. Not that anyone hath seen the Holiest at any time, save they which are of the Holiest, they alone, see the Holiest. Verily, verily, I say unto you, They who believe the Truth, have everlasting life.

LECTION 30. 8. -The original Gospels know nothing of the modern doctrine of the Anglican Church. Iesus was "the son of Mary and Joseph, whose parentage we know." This does not contradict, but rather suggests (to reconcile with Church doctrine), the Immaculate Conception of both parents to which the Church is now tending.

The Gospel of the Holy Twelve

Translated from the original Aramaic
by Rev. G.J.R. Ouseley

Section 4, Lections 31 thru 40

Lection 31

The Bread of Life And The Living Vine.
Iesus Rebuketh The Thoughtless Driver.

1. AGAIN Iesus said, I am the true Bread and the living Vine. Your fathers did eat manna in the wilderness and are dead. This is the food of God which cometh down from heaven, that whosoever eat thereof shall not die. I am the living food which came down from heaven, if any eat of this food they shall live for ever; and the bread that I will give is My truth and the wine which I will give is my life.

2. And the Jews strove amongst themselves, saying, How can this man give us himself for food? Then Iesus said, Think ye that I speak of the eating of flesh, which ye ignorantly do in the Temple of God?

3. Verily my body is the substance of God, and this is meat indeed, and my blood is the life of God and this is drink indeed. Not as your ancestors, who craved for flesh, and God gave them flesh in his wrath, and they ate of corruption till it stank in their nostrils, and their carcases fell by the thousand in the wilderness by reason of the plague.

4. Of such it is written, They shall wander nine and forty years in the wilderness till they are purified from their lusts, ere they enter into the land of rest, yea, seven times seven years shall they wander because they have not known My ways, neither obeyed My laws.

5. But They who eat this flesh and drink this blood dwell in me and I in them. As the Father-Mother of life hath sent me, and by Whom I live, so they that eat of me who am the truth and the life, even they shall live by me.

6. This is that living bread which coming down from heaven giveth life to the world. Not as your ancestors did eat manna and are dead. They that eat of this bread and this fruit, shall live for ever. These things said he in the synagogue, as he taught in Capernaum. Many therefore of his disciples, when they heard this, said, This is an hard saying, who can receive it?

7. When Iesus knew in himself that his disciples murmured at it, he said unto them, Doth this offend you? What and if ye shall see the Son and Daughter of man ascend to where they were before? It is the spirit that quickeneth, the flesh and blood profiteth nothing. The words that I speak unto you, they are spirit and they are life.

8. But there are some of you that believe not, For Iesus knew from the beginning who they were who should believe not, and who should betray him. Therefore said he unto them. No one can come unto me, except it were given from above.

9. From that time many of his disciples went back and walked no more with him. Then said Iesus unto the twelve, Will ye also go away?

10. Then Simon Peter answered him, Lord to whom shall we go? thou hast the words of eternal life. And we believe and we are sure that thou art that Christ, a Son of the living God.

11. Iesus answered them, Have not I chosen you Twelve, and one also who is a traitor? He spake of Judas Iscariot son of Simon the Levite, for he it was that should betray him.

12. AND Iesus was travelling to Jerusalem, and there came a camel heavy laden with wood. and the camel could not drag it up the hill whither he went for the weight thereof, and the driver beat him and cruelly ill-treated him, but he could make him go no further.

13. And Iesus seeing this, said unto him, Wherefore beatest thou thy brother? And the man answered, I wot not that he is my brother, is he not a beast of burden and made to serve me?

14. And Iesus said, Hath not the same God made of the same substance the camel and thy children who serve thee, and have ye not one breath of life which ye have both received from God?

15. And the man marvelled much at this saying, and he ceased from beating the camel, and took off some of the burden and the camel walked up the hill as Iesus went before him, and stopped no more till he ended his journey.

16. And the camel knew Iesus, having felt of the love of God in him. And the man inquired further of the doctrine, and Jesus taught him gladly and he became his disciple.

LECTION 31. 4. -Iesus quoted from a more ancient version than we now possess.

Lection 32

God the Food and Drink of All.

1. AND it came to pass as he sat at supper with his disciples one of them said unto him: Master, how sayest thou that thou wilt give thy flesh to eat and thy blood to drink, for it is a hard saying unto many?

2. And Jesus answered and said: The words which I spake unto you are Spirit and they are Life. To the ignorant and the carnally minded they savour of bloodshed and death, but blessed are they who understand.

3. Behold the corn which groweth up into ripeness and is cut down, and ground in the mill, and baked with fire into bread! of this bread is my body made, which ye see: and lo the grapes which grow on the vine unto ripeness, and are plucked and crushed in the winepress and yield the fruit of the vine! of this fruit of the vine and of water is made my blood.

4. For of the fruits of the trees and the seeds of the herbs alone do I partake, and these are changed by the Spirit into my flesh and my blood. Of these alone and their like shall ye eat who believe in me, and are my disciples, for of these, in the Spirit come life and health and healing unto man.

5. Verily shall my Presence be with you in the Substance and Life of God, manifested in this body, and this blood; and of these shall ye all eat and drink who believe in me.

6. For in all places I shall be lifted up for the life of the world, as it is written in the prophets; From the rising up of the sun unto the going down of the same, in every place a pure Oblation with incense shall be offered unto my Name.

7. As in the natural so in the spiritual. My doctrine and my life shall be meat and drink unto you, —the Bread of Life and the Wine of Salvation.

8. As the corn and the grapes are transmuted into flesh and blood, so must your natural minds be changed into spiritual. Seek ye the Transmutation of the natural into the Spiritual.

9. Verily I say unto you, in the beginning, all creatures of God did find their sustenance in the herbs and the fruits of the earth alone, till the ignorance and the selfishness of man turned many of them from the use which God had given them to that which was contrary to their original use, but even these shall yet return to their natural food, as it is written in the prophets, and their words shall not fail.

10. Verily God ever giveth of the Eternal Life and Substance to renew the forms of the universe. It is therefore of the flesh and blood, even the Substance and Life of the Eternal, that ye are partakers unto life, and my words are spirit and they are life.

11. And if ye keep My commandments and live the life of the righteous, happy shall ye be in this life, and in that which is to come. Marvel not therefore that I said unto you, Except ye eat of the flesh and drink the blood of God, ye have no life in you.

12. And the disciples answered saying: Lord, evermore give us to eat of this bread, and to drink of this cup, for thy words are meat and drink indeed;. By thy Life and by thy Substance may we live forever.

LECTION 32. 4,5, 8. -The true significance of the bread and the wine in the Holy Eucharist is here taught by anticipation -the substance and life of the Eternal One given and shed for the sustenance of the universe, and this does not exclude, but contains, all other mystical significations which piety suggests, as good, beautiful, and true -each in its place.

Lection 33

By The Shedding Of Blood Of Others Is No Remission Of Sins.

1. IESUS was teaching his disciples in the outer court of the Temple and one of them said unto him: Master, it is said by the priests that without shedding of blood there is no remission. Can then the blood offering of the law take away sin?

2. And Iesus answered: No blood offering, of beast or bird, or man, can take away sin, for how can the conscience be purged from sin by the shedding of innocent blood? Nay, it will increase the condemnation.

3. The priests indeed receive such offering as a reconciliation of the worshippers for the trespasses against the law of Moses, but for sins against the Law of God there can be no remission, save by repentance and amendment.

4. Is it not written in the prophets, Put your blood sacrifices to your burnt offerings, and away with them, and cease ye from the eating of flesh, for I spake not to your fathers nor commanded them, when I brought them out of Egypt, concerning these things? But this thing I commanded saying:

5, Obey my voice and walk in the ways that I have commanded you, and ye shall be my people, and it shall be well with you. But they hearkened not, nor inclined their ear.

6. And what doth the Eternal command you but to do justice, love mercy and walk humbly with your God? Is it not written that in the beginning God ordained the fruits of the trees and the seeds and the herbs to be food for all flesh?

7. But they have made the House of Prayer a den of thieves, and for the pure Oblation with Incense, they have polluted my altars with blood, and eaten of the flesh of the slain.

8. But I say unto you: Shed no innocent blood nor eat ye flesh. Walk uprightly, love mercy, and do justly, and your days shall be long in the land.

9. The corn that groweth from the earth with the other grain, is it not transmuted by the Spirit into my flesh? The grapes of the vineyard, with the other fruits are they not transmuted by the Spirit into my blood? Let these, with your bodies and souls be your Memorial to the Eternal.

10. In these is the presence of God manifest as the Substance and as the Life of the world. Of these shall ye eat and drink for the remission of sins, and for eternal life, to all who obey my words.

11. Now there is at Jerusalem by the sheep market, a pool which is called Bethesda, having five porches. In these lay a great multitude of impotent folk, of blind, halt, withered, waiting for the moving of the waters.

12. For at a certain season, an angel went down into the pool and troubled the waters; whosoever went first into the waters was made whole of whatever disease he had. And a man impotent from his birth was there.

13. And Iesus said unto him. Bring not the waters healing? He said unto him. Yea, Lord, but I have no man when the water is troubled to put me in, and while I am trying to come another steppeth down before me. And Jesus said to him, Arise, take up thy bed and walk. And immediately he rose and walked. And on the same day was the Sabbath.

14. The Jews therefore said to him, It is the Sabbath it is not lawful for thee to carry thy bed. And he that was healed wist not that it was Iesus. And Iesus had conveyed himself away, a multitude being in that place.

LECTION 33. 4. -Here is given the true significance of Ier. vii. 22, or as it should be rendered in that place, " Ye add burnt sacrifice to burnt offering and ye eat flesh. But I spake not to your fathers nor commanded them concerning these things," etc. Else, as translated in the A. V. it is inimical to the sense, see Numbers xi.

Lection 34

Love of Iesus for All Creatures.

1. WHEN Jesus knew how the Pharisees had murmured and complained because he made and baptized more disciples than John, he left Judea, and departed unto Galilee.

2. AND Jesus came to a certain Tree and abode beneath it many days. And there came Mary Magdalene and other women and ministered unto him of their substance, and he taught daily all that came unto him.

3. And the birds gathered around him, and welcomed him with their song, and other living creatures came unto his feet, and he fed them, and they ate out of his hands.

4. And when he departed he blessed the women who shewed love unto him, and turning to the fig tree, he blessed it also, saying. Thou hast given me shelter and shade from the burning heat, and withal thou hast given me food also.

5. Blessed be thou, increase and be fruitful, and let all who come to thee, find rest and shade and food, and let the birds of the air rejoice in thy branches.

6. And behold the tree grew and flourished exceedingly, and its branches took root downward, and sent shoots upward, and it spread mightily, so that no tree was like unto it for its size and beauty, and the abundance and goodness of its fruit.

7. AND as Jesus entered into a certain village he saw a young cat which had none to care for her, and she was hungry and cried unto him, and he took her up, and put her inside his garment, and she lay in his bosom.

8. And when he came into the village he set food and drink before the cat, and she ate and drank, and shewed thanks unto him. And he gave her unto one of his disciples, who was a widow, whose name was Lorenza, and she took care of her.

9. And some of the people said, This man careth for all creatures, are they his brothers and sisters that he should love them ? And he said unto them, Verily these are your fellow creatures of the great Household of God, yea, they are your brethren and sisters, having the same breath of life in the Eternal.

10. And whosoever careth for one of the least of these, and giveth it to eat and drink in its need, the same doeth it unto me, and whoso willingly suffereth one of these to be in want, and defendeth it not when evilly entreated, suffereth the evil as done unto me; for as ye have done in this life, so shall it be done unto you in the life to come.

LECTION 34. 2. -This beautiful incident does not stand alone in history, a similar story is related of Buddha, the Enlightener of India, the "Light of the East"; nor is it by any means irreverent to suppose that similar things should happen to persons of similar minds.

Lection 35

The Good Law. -Mary And Martha.

1. AND behold a certain lawyer stood up and tempted him, saying, Master, what shall I do to gain eternal life? He said unto him, What is written in the law ? how readest thou ?

2. And he answering, said, Thou shalt not do unto others, as thou wouldst not that they should do unto thee. Thou shalt love thy God with all thy heart and all thy soul and all thy mind. Thou shalt do unto others, as thou wouldst that they should do unto thee.

3. And he said unto him, Thou hast answered right, this do and thou shalt live; on these three commandments hang all the law and the prophets, for who loveth God, loveth his Neighbour also.

4. But he, willing to justify himself, said unto Jesus, And who is my neighbour? And Jesus answering said, A certain man

went down from Jerusalem to Jericho, and fell among, thieves, which stripped him of his raiment and wounded him and departed leaving him half dead.

5. And by chance there came down a certain priest that way, and when he saw him he passed by on the one side. And likewise a Levite also came and looked on him, and passed by on the other side.

6. But a certain Samaritan, as he journeyed, came where he was, and when he saw him he had compassion on him. And went to him and bound up his wounds, pouring in oil and wine, and set him on his own beast, and brought him to an inn and took care of him.

7. And on the morrow when he departed he took out two pence, and gave them to the host, and said, Take care of him and whatsoever thou spendest more, when I come again, I will repay thee.

8. Which now of these three, thinkest thou, was neighbour unto him that fell among thieves? And he said. He that shewed mercy on him. Then said Iesus unto him, Go, and do thou likewise.

9. Now it came to pass, as they went, that he entered into a certain village, and a woman named Martha received him into her house. And she had a sister called Mary, who also sat at Iesus, feet, and heard his word.

10. But Martha was cumbered about much serving and came to him saying, Lord, dost thou not care that my sister hath left me to serve alone? bid her therefore that she may help me.

11. And Iesus answered and said unto her, Martha, Martha, thou art careful and troubled about many things, but one thing is needful, and Mary hath chosen that good part, which shall not be taken away from her.

12. AGAIN, as Iesus sat at supper with his disciples in a certain city, he said unto them, As a Table set upon twelve pillars, so am I in the midst of you.

13. Verily I say unto you, Wisdom buildeth her house and heweth out her twelve pillars. She doth prepare her bread and her oil, and mingle her wine. She doth furnish her table.

14. And she standeth upon the high places of the city, and crieth to the sons and the daughters of men! Whosoever will, let them turn in hither, let them eat of my bread and take of my oil, and drink of my wine.

15. Forsake the foolish and live, and go in the way of understanding. The veneration of God is the beginning of wisdom, and the knowledge of the holy One is understanding. By me shall your days be multiplied, and the years of your life shall be increased.

LECTION 35. 2. -Although these words do not occur as they stand *verbatim* in any version of the Law of Moses as commonly received, the *spirit* of them certainly is there to be found, and in the original copy of the law (the best portion of which has been recovered by spiritual revelation) the very words also. And this original version was doubtless known to this young lawyer, as it evidently was to Iesus, when afterwards he gave the new law to his disciples on the holy Mount when he was transfigured before them in the company of Moses and Elias, the representatives of the old law, which was itself transfigured into the New.

v. 9. -" But one thing is needful " has been interpreted by some, not without reason or probability, as meaning that there were flesh and non-flesh food at the feast, and so he said to Martha, " but one thing (dishes or food) is needful, and Mary hath chosen the better portion." Meaning also, in the spiritual plane, the pure food of heavenly wisdom for the soul. It may have been spoken against luxurious multiplicity of dishes in general. (See Dr. A. Clarke i,1, *loco.*)

Lection 36

The Woman Taken In Adultery.

1. ON a certain day, early in the morning, Iesus came again into the temple, and all the people came unto him, and he sat down and taught them.

2. And the scribes and Pharisees brought unto him a woman taken in adultery, and when they had set her in the midst, they said unto him, Master, this woman was taken in adultery, in the very acts. Now Moses in the law commanded us that such should be stoned, but what sayest thou?

3. This they said, tempting him, that they might have to accuse him. But Iesus stooped down, and with his finger wrote on the ground, as though he heard them not.

4. So when they continued asking him, he lifted up himself, and said unto them, He that is without sin among you, let him cast the first stone at her.

5. And again he stooped down and wrote on the ground. And they which heard it, being convicted by their own conscience, went out one by one, beginning at the eldest, even unto the last; and Iesus was left alone, and the woman standing in the midst.

6. When Iesus had lifted up himself, and saw none but the woman, he said unto her, Woman, where are those thine accusers? hath no man condemned thee? She said unto him, No man, Lord. And Iesus said unto her, Neither do I condemn thee. From henceforth sin no more; go in peace.

7. AND he spake this parable unto certain which trusted in themselves that they were righteous, and despised others: Two men went up into the Temple to pray; the one a rich Pharisee, learned in the law, and the other a taxgatherer, who was a sinner.

8. The Pharisee stood and prayed thus with himself; God, I thank thee, that I am not as other men are, extortioners, unjust, adulterers, or even as this taxgatherer. I fast twice in the week, 1 give tithes of all that I possess,

9. And the taxgatherer, standing afar off, would not lift up so much as his eyes unto heaven, but smote upon his breath, saying, God be merciful to me a sinner.

10. I tell you, this man went down to his house justified rather than the other; for every one that exalteth himself shall be abased; and he that humbleth himself shall be exalted.

LECTION 36. 2-6. -This beautiful story, so characteristic of Iesus, has been most unjustifiably pronounced by modern revisers as an interpolation. It is a parable of human life, ever true, never old. "Let him who is without sin amongst you cast the first stone."

Lection 37

The Re-generation Of The Soul.

1. IESUS sat in the porch of the Temple, and some came to learn his doctrine, and one said unto him, Master, what teachest thou concerning life?
2. And he said unto them, Blessed are they who suffer many experiences, for they shall be made perfect through suffering: they shall be as the angels of God in Heaven and shall die no more, neither shall they be born any more, for death and birth have no more dominion over them.
3. They who have suffered and overcome shall be made Pillars in the Temple of my God, and they shall go out no more. Verily I say unto you, except ye be born again of water and of fire, ye cannot see the kingdom of God.
4. And a certain Rabbi (Nicodemus) came unto him by night for fear of the Jews, and said unto him, How can a man be born again when he is old? can he enter a second time into his mother's womb and be born again?
5. Iesus answered, Verily I say unto you except a man be born again of flesh and of spirit, he cannot enter into the kingdom of God. The wind bloweth where it listeth, and ye hear the sound thereof, but cannot tell whence it cometh or whither it goeth.
6. The light shineth from the East even unto the West; out of the darkness, the Sun ariseth and goeth down into darkness again; so is it with man, from the ages unto the ages.
7. When it cometh from the darkness, it is that he hath lived before, and when it goeth down again into darkness, it is that he may rest for a little, and thereafter again exist.
8. So through many changes must ye be made perfect, as it is written in the book of Job, I am a wanderer, changing place after place and house after house, until I come unto the City and Mansion which is eternal.
9. And Nicodemus said unto him, How can these things be? And Iesus answered and said unto him, Art thou a teacher in Israel, and understandeth not these things? Verily we speak that which we do know, and bear witness to that which we have seen, and ye receive not our witness.
10. If I have told you of earthly things and ye believe not, how shall ye believe if I tell you of Heavenly things? No man hath ascended into Heaven, but he that descended out of Heaven, even the Son-Daughter of man which is in Heaven.

LECTION 37. 8. -That our Lord spoke here primarily of a physical rebirth as the great aid of the spiritual re-birth, there can be no doubt, for he distinctly declares he had been telling Nicodemus of "earthly things" in the preceding words, albeit as the analogies and correspondences of spiritual things, as his usual method was. To interpret this dialogue, even as in the A. V., exclusively of the spiritual re-birth, is contrary to the plain meaning of the words.

Lection 38

Iesus Condemneth the Ill-Treatment Of Animals.

1. AND some of his disciples came and told him of a certain Egyptian, a son of Belial, who taught that it was lawful to torment animals, if their sufferings brought any profit to men.
2. And Iesus said unto them, Verily I say unto you, they who partake of benefits which are gotten by wronging one of God's creatures, cannot be righteous: nor can they touch holy things, or teach the mysteries of the kingdom, whose hands are stained With blood, or whose mouths are defiled with flesh.
3. God giveth the grains and the fruits of the earth for food: and for righteous man truly there is no other lawful sustenance for the body.
4. The robber who breaketh into the house made by man is guilty, but they who break into the house made by God, even of the least of these are the greater sinners. Wherefore I say unto all who desire to be my disciples, keep your hands from bloodshed and let no flesh meat enter your mouths, for God is just and bountiful, who ordaineth that man shall live by the fruits and seeds of the earth alone.
5. But if any animal suffer greatly, and if its life be a misery unto it. or if it be dangerous to you, release it from its life quickly, and with as little pain as you can, Send it forth in love and mercy, but torment it not, and God the Father-Mother will shew mercy unto you, as ye have shown mercy unto those given into your hands.

6. And whatsoever ye do unto the €ast of these my children, ye do it unto me. For I am in them and they are in me, Yea, I am in all creatures and all creatures are in me. In all their joys I rejoice, in all their afflictions I am afflicted. Wherefore I say unto you: Be ye kind one to another, and to all the creatures of God.

7. AND it came to pass the day after, that he came into a city called Nain; and many of his disciples went with him, and much people.

8. Now when he came nigh to the gate of the city, behold there was a dead man carried out the only son of his mother, and she was a widow: and much people of the city was with her.

9. And when the Lord saw her, he had compassion on her, and said unto her, Weep not, thy son sleepeth. And he came and touched the bier: and they that bare him stood still. And he said, Young man, I say unto thee, Arise.

10. And he that was esteemed dead sat up, and began to speak. And he delivered him to his mother. And there came an awe upon all: and they glorified God, saying, A great prophet is risen up among us; and God hath visited his people.

LECTION 38. -"Death" here, as in other cases, is a state of trance or suspended animation, not easily distinguishable from death even by the physician. A circumstance often leading to the revolting fact of burial alive -a fate, however, not so utterly hopeless in the East, where the dead are buried earth to earth in their shrouds, as in the countries of the West, with the modern and barbarous custom of closed coffins, with covers fastened down, and seven feet of earth over them. It is now ascertained by the more advanced and enlightened medical men, and others, and their official reports, that five per 1,000 must, in these English countries, come to this terrible fate, as there are yet no efforts made to prevent it, as in France, Holland, and other countries, where more rational and civilized practices prevail, and where it is found that five per 1,000 come to life, *before* actual interment, or show signs of premature burial after, when exhumed.

Lection 39

Seven Parables of the The Kingdom of Heaven.

1. AGAIN Iesus was sitting under the Fig tree, and his disciples gathered round him, and, round them came a multitude of people to hear him, and said unto them, Whereunto shall I liken the Kingdom of Heaven ?

2. AND he spake this parable, saying. The kingdom of Heaven is like to a certain seed, small among seeds, which a man taketh and soweth in his field, but when it is grown it becometh a great tree which sendeth forth its branches all around, which again, shooting downward into the earth take root and grow upward, till the field is covered by the tree, so that the birds of the air come and lodge in the branches thereof and the creatures of the earth find shelter beneath it.

3. ANOTHER parable put he forth unto them, saying, The kingdom of Heaven is like unto a great treasure hid in a field, the which when a man findeth he hideth it, and for joy thereof goeth and selleth all that he hath and buyeth that field, knowing how great will be the wealth therefrom,

4. AGAIN is the kingdom of Heaven like to one pearl of great price, which is found by a merchant seeking goodly pearls, and the merchant finding it, selleth all that he hath and buyeth it knowing how many more times it is worth than that which he gave for it.

5. AGAIN, the Kingdom of Heaven is like unto a woman who taketh of the incorruptible leaven and hideth it in three measures of meal, till the whole is leavened, and being baked by fire, becometh one loaf. Or, again, to one who taketh a measure of pure wine, and poureth it into two or four measures of water, till the whole being mingled becometh the fruit of the vine.

6. AGAIN, the Kingdom of Heaven is like unto a City built foursquare on the top of a high hill, and established on a rock, and strong in its surrounding wall, and its towers and its gates, which lie to the north, and to the south, and to the east, and to the west. Such a city falleth not, neither can it be hidden, and its gates are open unto all, who, having the keys, will enter therein.

7. AND he spake another parable, saying: The Kingdom of Heaven is like unto good seed that man sowed in his field, but in the night, while men slept, his enemy came and sowed tares also among the wheat, and went his way. But when the blade sprung up and brought forth fruit in the ear, there appeared the tares also.

8. And the servants of the householder came unto him and said, Sir, didst thou not sow good seed in thy field, whence then hath it tares? And he said unto them, An enemy hath done this.

9. And the servants said unto him, Wilt thou then that we go and gather them up ? But he said, Nay, lest haply while ye gather up the tares, ye root up the good wheat with them.

10. Let both grow together until the harvest, and in the time of the harvest I will say to the reapers,
Gather up first the tares and bind them in bundles to burn them and enrich the soil, but gather the wheat into my barn.

11. AND again he spake, saying, The kingdom of Heaven is like unto the sowing of seed. Behold a sower went forth to sow, and as he sowed, some seeds fell by the wayside, and the fowls of the air came and devoured them.

12. And others fell upon rocky places without much earth, and straightway they sprang up because they had no deepness of earth, and when the sun was risen they were scorched, and because they had no root they withered away.

13. And others fell among thorns, and the thorns grew up and choked them. And others fell upon good ground, ready prepared, and yielded fruit, some a hundredfold, some sixty, some thirty. They who have ears to hear let them hear.

Lection 40

Iesus Expounds His Inner Teaching To The Twelve.

1. AND the disciples came and said unto him, Why speakest thou unto the multitude in parables? He answered and said unto them, Because it is given unto you to know the mysteries of the kingdom of Heaven, but to them it is not given.

2. For whosoever hath to him shall be given and he shall have more abundance; but whosoever hath not, from him shall be taken away even that which he seemeth to have.

3. Therefore speak I to them in parables because they seeing see not, and hearing they hear not, neither do they understand.

4. For in them is fulfilled the prophecy of Esaias. which saith, Hearing ye shall hear and shall not understand and seeing ye shall see and shall not perceive; for this people's heart is waxed gross, and their ears are dull of hearing and their eyes they have closed, lest at any time they should see with their eyes, and hear with their ears, and should understand with their heart, and should be converted and I should heal them.

5. But blessed are your eyes for they see, and your ears for they hear, and your hearts for they understand. For verily I say unto you, That many prophets and righteous men have desired to see those things which ye See, and have not seen them, and hear those things which ye hear, and have not heard them.

6. THEN Iesus sent the multitude away and his disciples came unto him, saying, Declare unto us the parable of the field; and he answered and said unto them, He that soweth the good seed Is the Son of man; the field is the world, the good seed are the children of the kingdom, but the tares are the children of the wicked one. The enemy that sowed them is the devil, the harvest is the end of the world, and the reapers are the angels.

7. As therefore the tares are gathered and burned in the fire so shall it be in the end of this world. The Son of man shall send forth his angels, and they shall gather out of his kingdom all things that offend, and them which do iniquity, and shall cast them into a furnace of fire, and they who will not be purified shall be utterly consumed. Then shall the righteous shine forth as the Sun in the kingdom of Heaven.

8. HEAR ye also the parable of the sower. The seed that fell by the wayside is like as when any hear the word of the kingdom, and understand it not, then cometh the wicked one and catcheth away that which was sown in their heart. These are they which received seed by the wayside.

9. And they that received the seed into stony places, the same are they that hear the Word and anon with joy receive it. Yet have they not root in themselves but endure only a while, for when tribulation or persecution ariseth because of the Word, by and by they are offended.

10. They also that received seed among the thorns are they that hear the Word, and the cares of this world and the deceitfulness of riches choke the Word, and they become unfruitful.

11. But they that receive the seed into the good ground, are they that hear the Word and understand it, who also bear fruit and bring forth, some thirty, some sixty, some a hundred fold.

12. These things I declare unto you of the inner circle; but to those of the outer in parables. Let them hear who have ears to hear.

The Gospel of the Holy Twelve

Translated from the original Aramaic
by Rev. G.J.R. Ouseley

Section 5, Lections 41 thru 50

Lection 41

Iesus setteth free the Caged Birds.
The Blind Man who denied that others saw.

1. AND as Iesus was going to Iericho there met him a man with a cage full of birds which he had caught and some young doves. And he saw how they were in misery having lost their liberty, and moreover being tormented with hunger and thirst.

2. And he said unto the man, What doest thou with these? And the man answered, I go to make my living by selling these birds which I have taken.

3. And Iesus said, What thinkest thou, if another, stronger than thou or with greater craft, were to catch thee and bind thee, or thy wife, or thy children, and cast thee into a prison, in order to sell thee into captivity for his own profit, and to make a living?

4. Are not these thy fellow creatures, only weaker than thou? And doth not the same God our Father-Mother care for them as for thee? Let these thy little brethren and sisters go forth into freedom and see that thou do this thing no more, but provide honestly for thy living.

5. And the man marvelled at these words and at his authority, and he let the birds go free. So when the birds came forth they flew unto Jesus and stood on his shoulder and sang unto him.

6. And the man inquired further of his doctrine, and he went his way, and learnt the craft of making baskets, and by this craft he earned his bread, and afterwards he brake his cages and his traps, and became a disciple of Jesus.

7. AND Iesus beheld a man working on the Sabbath, and he said unto him, Man, if thou knowest not the law in the spirit; but if thou knowest not, thou art accursed and a transgresor of the law.

8. And again Iesus said unto his disciples, what shall be done unto these servants who, knowing their Lord's will, prepare not themselves for his coming, neither do according to his will?

9. Verily I say unto you, They that know their Master's will, and do it not, shall be beaten with many stripes. But they who not knowing their Master's will, do it not, shall be beaten with but few stripes. To whomsoever much is given, of them is much required. And to whom little is given from them is required but little.

10. AND there was a certain man who was blind from his birth. And he denied that there were such things as Sun, Moon, and Stars, or that colour existed. And they tried in vain to persuade him that other people saw them; and they led him to Iesus, and he anointed his eyes and made him to see.

11. And he greatly rejoiced with wonder and fear, and confessed that before he was blind. And now after this, he said, I see all, I know everything, I am god.

12. And Iesus again said unto him, How canst thou know all? Thou canst not see through the walls of thine house, nor read the thoughts of thy fellow men, nor understand the language of birds, or of beasts. Thou canst not even recall the events of thy former life, conception, or birth.

13. Remember with humility how much remains unknown to thee, yea, unseen, and doing so, thou mayest see more clearly.

LECTION 41. 11-13. -Very applicable to the present age also.

Lection 42

Iesus Teacheth Concerning Marriage The Blessing of Children

1. AND it came to pass that when Jesus had finished these sayings, he departed from Galilee and came into the coasts of Judea beyond Jordan; and great multitudes followed him and he healed them there.

2. The Pharisees also came unto him, tempting him and saying unto him, Is it lawful for a man to put away a wife for every cause?

3. And he answered and said unto them, In some nations, one man hath many wives, and putteth away whom he will for a just cause; and in some, a woman hath many husbands, and putteth away whom she will for a just cause; and in others, one man is joined to one woman, in mutual love, and this is the first and the better way.

4. For have ye not read that God who made them at the beginning, made them male and female, and said, For this cause shall a man or a woman leave father and mother, and shall cleave to his wife or her husband, and they twain shall be one flesh.

5. Wherefore they are no more twain, but one flesh. What therefore God have joined together, let not man put asunder.

6. They said unto him, Why did Moses then command to give a writing of divorcement? He saith unto them, Moses because of the hardness of your hearts suffered you to put away your wives. even as he permitted you to eat flesh, for many causes, but from the beginning it was not so.

7. And I say unto you, Whosoever shall put away a wife, except it be for a just cause, and shall marry another in her place, committeth adultery. His disciples say unto him, If the case of the man be so with his wife it is not good to marry.

8. But he said unto them All cannot receive this saying, save they to whom it is given. For there are some, celibates who were so born from their mother's womb, and there are some, which were made celibates of men, and there be some, who have made themselves celibates for the kingdom of Heaven's sake. He that is able to receive it, let him receive it.

9. THEN there came unto him little children that he should put his hands on them and bless them, and the disciples rebuked them.

10. But Jesus said, Suffer little children to come unto me and forbid them not, for of such is the kingdom of Heaven. And he laid his hands on them and blessed them.

11. AND as he entered into a certain village, there met him ten men that were lepers, which stood afar off. And they lifted up their voices, and said, Jesus Master, have mercy on us.

12. And when he saw them, he said unto them Go, shew yourselves unto the priests. And it came to pass, that, as they went, they were cleansed. And one of them, when he saw that he was healed, turned back, and with a loud voice glorified God and fell down on his face at his feet, giving him thanks: and he was a Samaritan.

13. And Jesus answering said, Were there not ten cleansed? but where are the nine? There are not found that returned to give glory to God, save this stranger. And he said unto him, Arise, go thy way: thy faith hath made thee whole.

Lection 43

Iesus Teacheth Concerning the Riches of this World and the Washing of Hands and Eating Of Unclean Meats

1. AND, behold, one came and said unto him. Good Master, what good thing shall I do, that I may have eternal life? And he said unto him, Why callest thou me good? there is none good but one, that is, God; but if thou wilt enter into life, keep the commandments. He saith unto him, which be they?

2. Iesus said, What teacheth Moses? Thou shalt not kill, thou shalt not commit adultery, thou shalt not steal, thou shalt not bear false witness, honor thy father and thy mother and thou shalt love thy neighbour

as thyself. The young man saith unto him, All these things have I kept from my youth up; what lack I yet?

3. Iesus said unto him, If thou wilt be perfect go and sell that thou hast in abundance, and give to those who have not, and thou shalt have treasure in heaven; and come and follow me.

4. But when the young man heard that saying, he went away sorrowful, for he had great possessions, yea, more than satisfied his needs.

5. Then said Iesus unto his disciple, Verily I say unto you, that the rich man shall hardly enter into the kingdom of Heaven. And again I say unto you, It is easier for a camel to go through the 'gate of the needle's eye" than for a rich person to enter into the kingdom of God.

6. When his disciples heard it, they were exceedingly amazed, saying, Who then can be saved? But Iesus beheld them, and said unto them, For the carnal mind this is impossible, but with the spiritual mind all things are possible.

7. And I say. unto you, Make not to yourselves friends of the Mammon of unrighteousness that when ye fail they may receive you into their earthly habitations; but rather of the true riches, even the Wisdom of God, that so ye may be received into everlasting mansions which fade not away.

8. Then Peter, said unto him, Behold we have forsaken all and followed thee. And Jesus said unto them, Verily I say unto you, that ye which have followed me, in the regeneration, when the Son of man shall sit in the throne of his glory, ye also shall sit upon twelve thrones, judging the twelve tribes of Israel, but the things of this world it is not mine to give.

9. And everyone that hath forsaken riches, houses, friends, for the kingdom of Heaven's sake and its righteousness, shall receive a hundred fold in the age to come and shall inherit everlasting life. But many that are first shall be last, and many that are last shall be first.

10. AND there came unto him certain of the Scribes and Pharisees who had seen one of his disciples eat with unwashed hands.

11. And they found fault, for the Jews eat not except they have first washen their hands and many other things observe they, in the washing of Cups and of vessels and of tables.

12. And they said, Why, walk not all thy disciples after the tradition of the elders, for we saw one who did eat with unwashed hands?

13. And Iesus said, Well hath Moses commanded you to be clean, and to keep your bodies clean, and your vessels clean, but ye have added things which ofttimes cannot be observed by every one at all times and in all places.

14. Hearken unto me therefore, not only unclean things entering into the body of man defile the man, but much more do evil thoughts and unclean, which pour from the heart of man, defile the inner man and defile others also. Therefore take heed to your thoughts and cleanse your hearts and let your food be pure.

15. These things ought ye to do, and not to leave the others undone. Whoso breaketh the law of purification of necessity, are blameless, for they do it not of their own will, neither despising the law which is just and good. For cleanliness in all things is great gain.

16. Be ye not followers of evil fashions of the world even in appearance; for many are led into evil by the outward seeming, and the likeness of evil.

Lection 44

The Confession of the Twelve.

1. AGAIN Iesus sat near the sea, in a circle of twelve palm trees, where he oft resorted, and the Twelve and their fellows came unto him, and they sat under the shade of the trees, and the holy One' taught them sitting in their midst.

2. And Iesus said unto them, Ye have heard what men in the world say concerning me, but whom do ye say that I am? Peter rose up with Andrew his brother and said, Thou art the Christ, the Son of the living God, who descendeth from heaven and dwelleth in the hearts of them who believe and obey unto righteousness. And the rest rose up and said, each after his own manner, These words are true, so we believe.

3. And Iesus answered them saying, Blessed are ye my twelve who believe, for flesh and blood hath not revealed this unto you, but the spirit of God which dwelleth in you. I indeed am the way, the Truth and the Life; and the Truth understandeth all things.

4. All truth is in God, and I bear witness unto the truth. I am the true Rock, and on this Rock do I build my Church, and the gates of Hades shall not prevail against it, and out of this Rock shall flow rivers of living water to give life to the peoples of the earth.

5. Ye are my chosen twelve. In me, the Head and Corner stone, are the twelve foundations of my house built on the rock, and on you in me shall my Church be built, and in truth and righteousness shall my Church be established.

6. And ye shall sit on twelve thrones and send forth light and truth to all the twelve tribes of Israel after the Spirit, and I will be with you, even unto the end of the world.

7. But there shall arise after you, men of perverse minds who shall through ignorance or through craft, suppress many things which I have spoken unto you, and lay to me things which I never taught, sowing tares among the good wheat which I have given you to sow in the world.

8. Then shall the truth of God endure the contradiction of sinners, for thus it hath been, and thus it will be. But the time cometh when the things which they have hidden shall be revealed and made known, and the truth shall make free those which were bound.

9. One is your Master, all ye are brethren, and one is not greater than another in the place which I have given unto you, for ye have one Master, even Christ, who is over you and with you and in you, and there is no inequality among my twelve, or their fellows.

10. All are equally near unto me. Strive ye not therefore for the first place, for ye are all first, because ye are the foundation stones and pillars of the Church, built on the truth which is in me and in you, and the truth and the law shall ye establish for all, as shall be given unto you.

11. Verily when ye and your fellows agree together touching anything in my Name, I am in the midst of you and with you.

12. Woe is the time when the spirit of the world entereth into the Church, and my doctrines and precepts are made void through the corruption of men and of women. Woe is the world when the Light is hidden. Woe is the world when these things shall be.

13. AT that time Iesus lifted his voice and said, I thank thee, O most righteous Parent, Creator of Heaven and Earth, that though these things are hidden from the wise and the prudent, they are nevertheless revealed unto babes.

14. No one knoweth thee, save the Son, who is the Daughter of man. None do know the Daughter or the Son save they to whom the Christ is revealed, who is the Two in One.

15. Come unto me all ye that labour and are heavy laden, and I will give you rest. Take my yoke upon you and learn of me, for I am meek and lowly in heart, and ye shall find rest unto your souls. For my yoke is equal and it is easy, my burden is light and presseth not unequally.

LECTION 44. 4. -In the Syrian Paschito, accepted by all Christians, in the Aramaic, the very language spoken by the Lord while on earth, the passage reads thus -"Thou art Kepha (rock), and on this rock I will build my Church, and the gates of Sheol shall not prevail against her." In the Aramaic original there is but one word for rock or stone, not two *petros* and *petra,* as in the Greek -on which anti-Catholic controversialists found their views. Rather do these words, "petros" and "petra," being simply the masculine and feminine forms, denote a *duality,* viz. Intuition and Intellects conjoined in the one human Individual; or in philosophy, the inductive and the deductive methods, the one of Plato, the other of Aristotle.

Lection 45

Seeking For Signs - The Unclean Spirit.

1. THEN certain of the Scribes and of the Pharisees answered saying, Master we would see a sign from thee. But he answered and said unto them, An evil and adulterous generation seeketh after a sign and there shall no sign be given to it, but the sign of the prophet Jonas.

2. Yea, as Jonas was three days and three nights in the whale's belly, so shall the Son of Man be three days and three nights in the heart of the earth, and after he shall rise again.

3. The men of Nineveh shall rise in judgment with this generation and shall condemn it, because they repented at the preaching of Jonas, and behold a greater than Jonas is here.

4. The Queen of the South shall rise up in the judgment with this generation, and shall condemn it; for she came from the uttermost parts of the earth to hear the wisdom of Solomon, and behold, a greater than Solomon is here.

5. AGAIN he said: When the unclean spirit is gone out of any, he walketh through dry places seeking rest, and finding none it saith, I will return into my house from whence I came out. And when he is come he findeth it empty, swept and garnished, for they asked not the Good Spirit to dwell within them, and be their eternal Guest.

6. Then he goeth and taketh with him seven other spirits more wicked than himself, and they enter in and dwell there, and the last state of all such is worse than the first. Even so shall it be also unto this wicked generation, which refuseth entrance to the Spirit of God.

7. For I say unto you, whosoever blasphemeth the Son of Man, it shall be forgiven them; but whoso blasphemeth the Holy Spirit it shall not be forgiven them either in this age, or in the next, for they resist the Light of God, by the false traditions of men.

8. WHILE, he yet talked to the people, behold his parents and his brethren and his sisters stood without, desiring to speak with him. Then one said unto him, Behold thy father and thy mother, and thy brethren and thy sisters stand without, desiring to speak with thee.

9. But he answered and said unto him that told him; Who is my father and who is my mother? and who are my brethren and my sisters?

10. And he stretched forth his hand towards his disciples and said, Behold my father and my mother, my brethren and sisters, and my children! For whosoever shall do the will Of my Parent Who is in Heaven the same is my father and my mother, my brother and my sister, my son and my daughter.

11. AND there were some Pharisees, who were covetous and proud of their riches, and he said unto them, Take heed unto yourselves, and beware of covetousness, for a man's life consisteth not in the abundance of things which he possesseth.

12. And he spake a parable unto them, saying, The ground of a certain rich man brought forth plentifully; and he thought within himself, saying, What shall I do, because I have no room where to bestow my fruits?

13. And he said, This will I do; I will pull down my barns, and build greater; and there will I bestow all my fruits and my goods.

14. And I will say to my soul, thou hast much goods laid up for many years, take thine ease, drink and be merry.

15. But God said unto him, Thou fool, this night thy life shall be required of thee; then whose shall those things be, which thou hast provided?

16. So are they that lay up treasures for themselves, and are not rich in good works to them that need, and are in want.

LECTION 45. 7. -The word "Blasphemy" is derived from the Greek , "blapto" to hinder, retard, obstruct (progress) and "pheme" to speak, thus signifying what is usually known as that conservatism which injures progress and obstructs it. Such a mind is held in bonds by too great deference to authority, or customs, or fashions, or traditions, and closes itself against fresh truths, or the restatement or development of old truths, and thus resists the light which may come through a divine messenger, and commits the "sin against the Holy Spirit" by closing eyes and ears against the higher teaching of God's truth. Such cannot be forgiven *i.e.* obtain release in this age (incarnation} or the succeeding. For the effects of persistent resistance are age enduring, like a prison house of the mind, and it cannot break the strong walls of prejudice and come forth, being fast bound, and needing, it may be ages, for its emancipation from error long cherished.

LECTION 45. 1O. -Without slighting any believer's love to the B.V.M., these words clearly shew how Iesus regarded *Righteousness* as relating all true believers to him more than any blood relationship could possibly do.

Lection 46

The Transfiguration on the Mount.
The Giving of the Law.

1. AFTER six days, when the Feast of Tabernacles was nigh at hand, Iesus taketh the twelve and bringeth them up into a high mountain apart, and as he was praying the fashion of his countenance was changed, and he was transfigured before them, and his face did shine as the sun, and his raiment was white as the light.

2. And, behold, there appeared unto them Moses and Elias talking with him and spake of the Law, and of his decease which he should accomplish at Jerusalem.

3.. And Moses spake, saying, This is he of whom I foretold, saying, A prophet from the midst of thy brethren, like unto me shall the Eternal send unto you, and that which the Eternal speaketh unto him, shall he speak unto you, and unto him shall ye hearken, and whoso will not obey shall bring upon themselves their own destruction.

4. Then Peter said unto Iesus, Lord, it is good for us to be here; if thou wilt let us make here three tabernacles; one for thee, and one for Moses, and one for Elias.

5. While he yet spake, behold a bright cloud overshadowed them, and twelve rays as of the sun issued from behind the cloud, and a voice came out of the cloud, which said, This is my beloved Son, in whom I am well pleased; hear ye him.

6. And when the disciples heard it, they fell on their faces and were sore amazed, and Jesus came and touched them and said, Arise and be not afraid. And when they had lifted up their eyes, they saw no man, save Jesus only. And the six glories were seen upon him.

7. AND Iesus said unto them, Behold a new law I give unto you, which is not new but old. Even as Moses gave the Ten Commandments to Israel after the flesh, so also I give unto you the Twelve for the Kingdom of Israel after the Spirit.

8. For who are the Israel of God ? Even they of every nation and tribe who work righteousness, love mercy and keep my commandments, these are the true Israel of God. And standing upon his feet, Jesus spake, saying:

9. Hear O Israel, JOVA, thy God is One; many are My seers, and My prophets. In Me all live and move, and have subsistence.

10. Ye shall not take away the life of any creature for your pleasure, nor for your profit. nor yet torment it.

11. Ye shall not steal the goods of any, nor gather lands and riches to yourselves, beyond your need or use.

12. Ye shall not eat the flesh, nor drink the blood of any slaughtered creature, nor yet any thing which bringeth disorder to your health or senses.

13. Ye shall not make impure marriages, where love and health are not, nor yet corrupt yourselves, or any creature made pure by the Holy.

14. Ye shall not bear false witness against any, nor wilfully deceive any by a lie to hurt them.

15. Ye shall not do unto others, as ye would not that others should do unto you.

16. Ye shall worship One Eternal, the Father-Mother in Heaven, of Whom are all things, and reverence the holy Name.

17. Ye shall revere your fathers and your mothers on earth, whose care is for you, and all the Teachers of Righteousness.

18. Ye shall cherish and protect the weak, and those who are oppressed, and all creatures that suffer wrong.

19. Ye shall work with your hands the things that are good and seemly; so shalt ye eat the fruits Of the earth, and live long in the land.

20. Ye shall purify yourselves daily and rest the Seventh Day from labour, keeping holy the Sabbaths and the Festival of your God.

21. Ye shall do unto others as ye would that others should do unto you.

22. And when the disciples heard these words, they smote upon their breasts, saying: Wherein we have offended. O God forgive us: and may thy wisdom, love and truth within us incline our hearts to love and keen this Holy Law.

23. And Jesus said unto them, My yoke is equal and my burden light, if ye will to bear it, to you it will be easy. Lay no other burden on those that enter into the kingdom, but only these necessary things.

24. This is the new Law unto the Israel of God, and the Law is within, for it is the Law of Love, and it is not new but old. Take heed that ye add nothing to this law, neither take anything from it. Verily I say unto you, they who believe and obey this law shall be saved, and they who know and obey it not, shall be lost.

25. But as in Adam all die so in Christ shall all be made alive. And the disobedient shall be purged

through many fires; and they who persist shall descend and shall perish eternally.

26. And as they came down from the mountain, Jesus charged them, saying, Tell the vision to no man, until the Son of man be risen again from the dead.

27. His disciples asked him, saying, Why then say the scribes that Elias must first come? And Jesus answered and said unto them, Elias truly shall first come and restore all things.

28. But I say unto you, that Elias is come already, and they knew him not, but have done unto him whatsoever they listed. Likewise shall also the Son of man suffer of them. Then the disciples understood that he spake unto them of John the Baptist.

LECTION 46. 2-7. -The appearance of Moses, to hand over, as it were, the Law and the Dispensation to Iesus, throws a light on the reason of the Transfiguration, which is lost in the accepted version. Elias also appears, so as to make, with Moses, the "two witnesses" required by the law. They recognize Iesus, and witness to him as the great Prophet whom God should raise to succeed and take the place of Moses as the Legislator of the New Dispensation.

The "six glories" may have reference to the six precepts In each table of the law as now given, or more probably to the six attributes of each of the two aspects of the Christ, six feminine and six masculine, as inherent in the Two in One -the " Father-Mother of the age to come." They are referred to in one of the most ancient gospels.

LECTION 46. 10-12 -Compare this with the Law as given by Moses in the "Book of the Going Forth of Israel," p. 35, all chiefly negative precepts. In the law given by Christ there are six negative and six positive. The negative is the external Way, in which certain actions are forbidden. The positive the interior way, in which certain duties are enjoined. As the prohibitions are summed up in the negative form of the Golden Rule, so the commands are summed up in the positive form. Had this law been faithfully observed by all Christians, there would have been no divisions or war as now, and this earth would have been a paradise for all. The allegations brought against Christianity and Christ as the cause of strife and bloodshed, are therefore baseless. It is the spirit of selfishness, of perversity, in direct opposition to the Law of Christ, which has been the root of all the evils laid to the charge of Christianity, as he said, "Behold I come to bring peace upon earth, but what if a sword cometh?" Besides the loss of life by wars and intemperance of all kinds, the amount of material wealth wasted yearly in Christian countries, is not only needless, but injurious things, forbidden by the Laws of Christ, is simply appalling. **More**

v. 12. -" Slaughtered " is the truer rendering which has in former editions been wrongly translated "living."

LECTION 46. 24..-The law as given through Moses "by the ministry of angels," and as the Church of Israel (the typical nation) received it through their Interpreters and Scribes, had no precept forbidding cruelty, oppression, flesh eating, drunkenness, the worship of mammon, impure marriages for money or position, and other vices and crimes which are the cause of nearly all the misery which now afflicts men and women " called to be happy sons and daughters of God Almighty , and which are visited by the holy Law on the children to the third and fourth generation of them that despise it, shewing mercy to all who love it." Nor yet did it include that mutual consideration for each other (positively and negatively) which is summed up in the one commandment of Christ " Love ye one another," not excluding therefrom the creatures which God hath given to be our earth mates and companions. **More**

Lection 47

The Spirit Giveth Life.
The Rich Man and the Beggar.

1. AND when they were come down from the Mount one of his disciples asked him, Master, if a man keep not all these commandments shall he enter into Life? And he said, the Law is good in the letter without the spirit is dead, but the spirit maketh the letter alive.

2. Take ye heed that ye obey from the heart, and in the spirit of love, all the Commandments which I have given unto you.

3. It hath been written, Thou shalt not kill, but I say unto you, if any hate and desire to slay, they are guilty of the law, yea, if they cause hurt or torture to any Innocent Creature they are guilty, But if they kill to put an end to suffering which cannot be healed, they are not guilty, if they do it quickly and in love.

4. It hath been said, Thou shalt not steal, but I say unto you, if any, not content with that which they have, desire and seek after that which is another's or if they withhold that which is just from the worker, they have stolen in their heart already, and their guilt is greater than that of one who stealeth a loaf in necessity, to satisfy his hunger.

5. Again ye have been told, Thou shalt not commit adultery, but I say unto you, if man or woman join together in marriage with unhealthy bodies, and beget unhealthy offspring, they are guilty, even though they have not taken their neighbour's spouse: and if any have not taken a woman who belongeth to another, but desire in their heart and seek after her, they have committed adultery already in spirit.

6. And again I say unto you, if any desire and seek to possess the body of any creature for food, or for pleasure, or for profit, they defile themselves thereby.

7. Yea, and if a man telleth the truth to his neighbour in such wise as to lead him into evil, even thought it be true in the letter, he is guilty.

8. Walk ye in the spirit, and thus shall ye fulfill the law and be meet for the kingdom. Let the Law be within your own hearts rather than on tables of memorial; which things nevertheless ye ought to do and not to leave the other undone for the Law which I have given unto you is holy, just and good, and blessed are all they who obey and walk therein.

9. God is Spirit, and they who worship God must worship in spirit and in truth, at all times, and in all places.

10. AND he spake this parable unto them who were rich, saying, There was a certain rich man, which was clothed in purple, and fine linen, and fared sumptuously every day.

11. And there was a certain beggar named Lazarus, which was laid at his gate, full of sores. And desiring to be fed with the crumbs which fell from the rich man's table; moreover the dogs came and licked his sores.

12. And it came to pass, that the beggar died, and was carried by the angels into Abraham's bosom; the rich man also died, and was buried with great pomp. And in Hades he lift up his eyes, being in torments, and seeth Abraham afar off, and Lazarus in his bosom.

13. And he cried and said, Father Abraham, have mercy on me, and send Lazarus, that he may dip the tip of his finger in water, and cool my tongue, for I am tormented in this place.

14. But Abraham said, Son, remember that thou in thy lifetime receivedst thy good things, and likewise Lazarus evil things: but now he is comforted, and thou art tormented. And thus are the changes of life for the perfecting of souls. And beside all this, between us and you there is a great gulf fixed, so that they which would pass from hence to you cannot; neither can they pass to us, that would come from thence, till their time be accomplished.

15. Then he said, I pray thee therefore, father, that thou wouldest send him to my Father's house; for I have five brethren, that he may testify unto them, lest they also came into this place of torment.

16. Abraham saith unto him, They have Moses and the prophets; let them hear them. And he said, Nay, father Abraham; but if one went unto them from the dead, they will repent.

17. And Abraham said unto him, if they hear not Moses and the prophets, neither will they be persuaded, though one rose from the dead.

LECTION 47. 2-7. -There is hardly a law that has been given by man or by God, but has been abused and turned into an instrument of oppression by the perversity of man, evading the spirit and insisting on the letter alone.

vv. 10-16. -This parable of Dives and Lazarus is pregnant with deepest teaching, utterly setting aside the narrow dogmas of the sects which have been founded on the traditional interpretation of it in the A. V.

Lection 48

Iesus Feedeth One Thousand with Five Melons.
Heals the Withered Hand On The Sabbath Day.

1. AND it came to pass as Iesus had been teaching the multitudes, and they were hungry and faint by reason of the heat of the day, that there passed by that way a woman on a camel laden with melons and other fruits.

2. And Iesus lifted up his voice and cried, O ye that thirst, seek ye the living water which cometh from Heaven, for this is the water of life, which whoso drinketh thirsteth not again.

3. And he took of the fruit, five melons and divided them among the people, and they eat, and their thirst was quenched, and he said unto them, If God maketh the sun to shine, and the water to fill out these fruits of the earth, shall not the Same be the Sun of your souls, and fill you with the water of life?

4. Seek ye the truth and let your souls be satisfied. The truth of God is that water which cometh from heaven, without money and without price, and they who drink shall be satisfied. And those whom he fed were one thousand men, women and children—and none of them went home a hungered or athirst; and many that had fever were healed.

5. At that time Iesus went on the Sabbath day through the cornfields, and his disciples were an hungered, and began to pluck the ears of corn, and to eat.

6. But when the Pharisees saw it, they said unto him, Behold, thy disciples do that which is not lawful to do upon the Sabbath day.

7. And he said unto them, Have ye not read what David did, when he was an hungered and they that were with him; how he entered into the house of God and did eat the shewbread, which was not lawful for him to eat, neither for them which were with him, but only for the priests?

8. Or have yet not read in the law, how that on the Sabbath days the priests in the Temple do work on the Sabbath and are blameless? But I say unto you, That in this place is One greater than the Temple.

9. But if ye had known what this meaneth, I will have mercy and not sacrifice, ye would not have condemned the guiltless. For the Son of man is Lord even of the Sabbath.

10. AND when he was departed thence, he went into their synagogue. And, behold, there was a man which had his hand withered. And they asked him, saying, is it lawful to heal on the Sabbath days? that they might accuse him.

11. And he said unto them, What man shall there be among you that shall have but one sheep, and if it fall into a pit on the Sabbath day will he not lay hold on it and lift it out? And if ye give help to a sheep, shall ye not also to a man that needeth?

12. Wherefore it is lawful to do well on the Sabbath day. Then saith he to the man, Stretch forth thine hand. And he stretched it forth, and it was restored whole, like as the other.

13. Then the Pharisees went out and held a council against him, how they might destroy him. But when Iesus knew it, he withdrew himself from thence; and great multitudes followed him, and he healed their sick and infirm, and charged them that they should not make it known.

14. So it was fulfilled, which was spoken by Esaias the prophet, saying, Behold my servant, whom I have chosen; my beloved, in whom my soul is well pleased; I will put my spirit upon him and he shall shew judgment to the Gentiles.

15. He shall not strive nor cry, neither shall any man hear his voice in the streets. A bruised reed shall he not break, and smoking flax shall he not quench till he send forth judgment unto victory. And in his Name shall the Gentiles trust.

LECTION 48. l-3. -The feeding of one thousand with five melons is another instance of the love of Iesus to the hungry and thirsty thousand who came to his ministry. He first feeds the body, then instructs the soul. Too often it is the reverse of this among his followers. Hence, "The lack of love causeth many to wax cold."

Lection 49

The True Temple of God

1. AND the Feast of the Passover was at hand. And it came to pass that some of the disciples being masons, were set to repair one of the chambers Of the Temple. And Iesus was passing by, and they said unto him, Master, Sees't thou these great buildings and what manner of stones are here, and how beautiful is the work of our ancestors?

2. And Iesus said, Yea, it is beautiful and well wrought are the stones, but the time cometh when not one stone shall be left on another, for the enemy shall overthrow both the city and the Temple.

3. But the true Temple is the body of man in which God dwelleth by the Spirit, and when this Temple is destroyed, in three days, God raiseth up a more glorious temple, which the eye of the natural man perceiveth not.

4. Know ye not that ye are the temples of the holy spirit? and whoso destroyeth one of these temples the same shall be himself destroyed.

5. AND some or the scribes, hearing him, sought to entangle him in his talk and said, If thou wouldst put away the sacrifices of sheep and oxen and birds, to what purpose was this Temple built for God by Solomon, which has been now forty and six years in restoring?

6. And Iesus answered and said, It is written in the prophets, My house shall be called a house of prayer for all nations, for the sacrifice of praise and thanksgiving. But ye have made it a house of slaughter and filled it with abominations.

7. Again it is written, From the rising of the sun unto the setting of the same, my Name shall be great among the Gentiles, and incense with a pure Offering shall be offered unto me. But ye have made it a desolation with your offerings of blood and used the sweet incense only to cover the ill savor thereof. I am come not to destroy the law but to fulfill it.

8. Know ye not what is written? Obedience is better than sacrifice and to hearken than the fat of rams. I, the Lord, am weary of your burnt offerings, and vain oblations, your hands are full of blood.

9. And is it not written, what is the true sacrifice? Wash you and make you clean and put away the evil from before mine eyes, cease to do evil, learn to do well. Do justice for the fatherless and the widow and all that are oppressed. So doing ye shall fulfill the law.

10. The day cometh when all that which is in the outer court, which pertaineth to blood offerings, shall be taken away and pure worshippers shall worship the Eternal in purity and in truth.

11. And they said, Who art thou that seekest to do away with the sacrifices, and despiseth the seed of Abraham? From the Greeks and the Egyptians hast thou learnt this blasphemy?

12. And Iesus said, Before Abraham was, I Am. And they refused to listen and some said, he is inspired by a demon, and others said, he is mad; and they went their way and told these things to the priests and elders. And they were wrath, saying, He hath spoken blasphemy.

LECTION 49. 7. -Iesus here disclaims the sacrifices of bleeding victims as being by part of true religion, but rather contrary thereto, adding to the guilt of the offerers. How often religion, so called, is put before morality, and dogma before loving-kindness and mercy.

Lection 50

Christ the Light of the World.

1. THEN spake Iesus again unto them, saying, I am the Light of the world: he that followeth me shall not walk in darkness, but shall have the light of life.

2. The Pharisees therefore said unto him, Thou bearest record of thyself thy record is not true.

3. Iesus answered and said unto them, Though I bear record of myself, yet my record is true: for I know whence I came, and whither I go: but ye cannot tell whence I come, and whither I go.

4. Ye judge after the flesh; I judge no man. And yet if I judge, my judgment is true: for I am not alone, but I come from the Father-Mother who sent me.

5. It is also written in your law, that the testimony of two men is true. I am one that bear witness of myself, Iohn bore witness of me, and he is a prophet, and the Spirit of truth that sent me bareth witness of me.

6. Then said they unto him, Where is thy Father and thy Mother? Iesus answered, Ye neither know me, nor my Parent: if ye had known me, ye should have known my Father and my Mother also.

7. And one said, shew us the Father, shew us the Mother, and we will believe thee. And he answered saying, if thou hast seen thy brother and felt his love, thou hast seen the Father, if thou hast seen thy sister and felt her love thou hast seen the Mother.

8. Far and near, the All Holy knoweth Their own, yea, in each of you, the Fatherhood and the

Motherhood may be seen, for the Father and the Mother are One in God.

9. These words spake Iesus in the treasury, as he taught in the temple. And no man laid hands on him; for his hour was not yet come. Then said Iesus again unto them, I go my way, and ye shall seek me, and shall die in your sins; whither I go, ye cannot come.

10. Then said the Jews, Will he kill himself? because he said, Whither I go, ye cannot come. And he said unto them, Ye are from beneath; I am from above; ye are of this world; I am not of this world.

11. I said therefore unto you, that ye shall die in your sins; for if ye believe not that I Am of God, ye shall die in your sins.

12. Then said they unto him, Who art thou? And Jesus said unto them, Even the Same that I said unto you from the beginning.

13. I have many things to say which shall judge you: but the Holy One that sent me is true; and I speak to the world those things which I have heard from above.

14. Then said Iesus unto them, When ye have lifted up the Son of man, then shall ye know that I am sent of God, and that I do nothing of myself; but as the All Holy hath taught me, I speak these things. Who sent me is with me: the All Holy hath not left me alone; for I do always those things that please the Eternal.

15. As he spake these words, many believed on him, for they said, He is a Prophet sent from God. Him let us hear.

The Gospel of the Holy Twelve

Translated from the original Aramaic
by Rev. G.J.R. Ouseley

Section 6, Lections 51 thru 60

Lection 51

The Truth Maketh Free

1. THEN said Jesus to those Jews which believed on him, If ye continue in my word, then are ye my disciples indeed; And ye shall know the truth, and the truth shall make you free.

2. They answered him, We be Abraham's seed, and were never in bondage to any man: how sayest thou, Ye shall be made free? Iesus answered them Verily, verily, I say unto you, Whosoever committeth sin is the servant of sin. And the servant abideth not in the house for ever: but the Son even the Daughter abideth ever.

3. If the Son therefore shall make you free, ye shall be free indeed. I know that ye are Abraham's seed after the flesh; but ye seek to kill me, because my word hath no place in you.

4. I speak that which I have seen with my Parent and ye do that which ye have seen with your parent. They answered and said unto him, Abraham is our father. Iesus said unto them, If ye were Abraham's children, ye would do the works of Abraham.

5. But now ye seek to kill me, a man that hath told you the truth, which I have heard of God: this did not Abraham. YE do the deeds of your father. Then said they to him, We be not born of fornication; we have one Father, even God.

6. Iesus said unto them, If God were your Parent, ye would love me: for I proceeded forth and came from God; neither came I of myself, but the All Holy sent me. Why do ye not understand my speech? even because ye cannot hear my word.

7. Ye are of your father the devil, and the lusts of your father ye will do. He was a murderer from the beginning, and abode not in the truth, because there is no truth in him.

8. When he speaketh a lie, he speaketh of his own; for he is a liar, and the father of it. And because I tell you the truth, ye believe me not.

9. As Moses lifted up the Serpent in the wilderness, so must the Son and Daughter of man be lifted up, that whosoever gazeth, believing should not perish, but have everlasting life.

10. Which of you convicteth me of sin ? And if I say, the truth, why do ye not believe me? He that is of God heareth God's words: ye therefore hear them not, because ye are not of God.

11. Then answered the Jews, and said unto him, Say we not well that thou art a Samaritan, and hath a demon ? Iesus answered, I have not a demon; but I honour the All Holy, and ye do dishonour me. And I seek not mine own glory, but the glory of God. But there is One who judgeth.

12. And certain of the Elders and Scribes from the Temple came unto him saying, Why do thy disciples teach men that it is unlawful to eat the flesh of beasts though they be offered in sacrifice as by Moses ordained.

13. For it is written, God said to Noah, The fear and the dread of you shall be upon every beast of the field, and every bird of the air, and every fish of the sea, into your hand they are delivered.

14. And Jesus said unto them, Ye hypocrites, well did Esaias speak of you, and your forefathers, sayings This people draweth nigh unto Me, with their mouths, and honour me with their lips, but their heart is far from me, for in vain do they worship Me teaching and believing, and teaching for divine doctrines, the commandments of men in my name but to satisfy their own lusts.

15. As also Jeremiah bear witness when he saith, concerning blood offerings and sacrifices I the Lord God commanded none of these things in the day that ye came out of Egypt, but only this I commanded you to do, righteousness, walk in the ancient paths, do justice, love mercy, and walk humbly with thy God.

16. But ye did not hearken to Me, Who in the beginning gave you all manner of seed, and fruit of the trees and seed having been for the food and healing of man and beast. And they said, Thou speakest against the law.

17. And he said against Moses indeed I do not speak nor against the law, but against them who corrupted his law, which he permitted for the hardness of your hearts.

18. But, behold, a greater than Moses is here! and they were wrath and took up stones to cast at him. And Jesus passed through their midst and was hidden from their violence.

LECTION 51. 2. -Many people think that perfect freedom is the power to do wrong as well as right. Such is not Christ's teaching. Freedom is the power to do moral good, nothing else : the other is not freedom, but slavery to the evil nature. This is the teaching of Rosmini and of the Franciscans, and is evidently the teaching of Christ. Other animals than man have the freedom essential to their nature, which, if they are allowed to follow, is a kind of moral good. In a true state of nature, the other animals are found innocent, till corrupted by the cruelty of man. **Cont.**

Lection 52

He Declareth His Pre-Existence.

1. ANOTHER time Iesus said, Verily, verily, I say unto you, If a man keep my saying, he shall never see death. Then said the Jews unto him, Now we know that thou hast a demon.

2. Abraham is dead, and the prophets; and thou sayest, If a man keep my saying, he shall never taste of death. Art thou greater than our father Abraham, which is dead ? and the Prophets are dead: whom makest thou thyself ?

3. Iesus answered, If I honour myself, my honour is nothing: it is my Father that honoureth me; of whom ye say, that he is your God: Yet ye have not known him; but I now him: and if I should say I know him not I shall be a liar like unto you; but I know the All Holy and am known of the Eternal.

4. Your father Abraham rejoiced to see my day; and he saw it, and was glad. Then said the Jews unto him, Thou art not yet forty five years old, and hast thou seen Abraham?

5. Iesus said unto them, Verily, verily, I say unto you, Before Abraham was, I AM.

6. And he said unto them, The All Holy hath sent you many prophets, but ye rose against them that were contrary to your lusts, reviling some and slaying others.

7. Then took they up stones to cast at him: but Iesus was hidden, and went out of the temple, through the midst of them, and so again passed unseen by them.

8. Again when his disciples were with him in a place apart, one of them asked him concerning the kingdom, and he said unto them:

9. As it is above, so it is below. As it is within, so it is without. As on the right hand, so on the left. As it is before, so it is behind. As with the great so with the small. As with the male, so with the female. When these things shall be seen, then ye shall see the kingdom of God.

10. For in me there is neither Male nor Female, but both are One in the All perfect. The woman is not without the man, nor is the man without the woman.

11. Wisdom is not without love, nor is love without wisdom. The head is not without the heart, nor is the heart without the head, in the Christ who atoneth all things. For God hath made all things by number, by weight, and by measure, corresponding, the one with the other.

12. These things are for them that understand, to believe. If they understand not, they are not for them. For to believe is to understand, and to believe not, is not to understand.

LECTION 52. 4. -The testimony of those who saw and knew Jesus as to his age, has been strangely ignored by writers of Biblical history and by the Church in general. This matter is briefly discussed elsewhere in these Notes, and deserves the attention of every student and thoughtful person. (See Notes liv.14-16; xcv.9.)

Lection 53

Iesus Healeth The Blind On The Sabbath.

1. AND at another time as Iesus passed by, he saw a man which was blind from his birth. And his disciples asked him saying, **Master, who did sin, this man, or his parents, that he was born blind?**

2. Iesus answered, To what purport is it, whether this man sinned, or his parents, so that the works of God are made manifest in him? I must work the works of my Parent who sent me, while it is day; the night cometh, when no man can work. As long as I am in the world, I am the Light of the world.

3. When he had thus spoken, he spat on the ground, and mingled clay with the spittle, and he anointed the eyes of the blind man with the clay And said unto him, Go, wash in the pool of Siloam (this meaneth by interpretation, Sent.) He went his way therefore, and washed, and came seeing.

4. The neighbours therefore, and they which before had seen him that he was blind, said, Is not this he that sat and begged? Some said, This is he: others said, He is like him: but he said, I am he.

5. Therefore said they unto him, How were thine eyes opened? He answered and said, A man that is called Jesus made clay, and anointed mine eyes, and said unto me, Go to the pool of Siloam, and wash: and I went and washed, and I received sight.

6. Then said they unto him, Where is he? He said, I know not where he is, that made me whole.

7. Then came to Him certain of the Sadducces, who deny that there is a resurrection, and they asked him saying, Master, Moses wrote unto us, if any man's brother die having a wife and leaving no children, that his brother should take his wife and raise up seed to his brother.

8. Now there were six brethren, and the first took a wife and he died childless: And the second took her to wife and he died childless: And the third, even unto the sixth, and they died also leaving no children Last of all the woman died also.

9. Now in the resurrection, whose of them is she, for the six had her to wife.

10 And Iesus answered them saying, whether a woman with six husbands, or a man with six wives, the case is the same. For the children of this world marry and are given in marriage.

11. But they, which being worthy, attain to the resurrection from the dead, neither marry, nor are given in marriage, neither can they die any more, for they are equal to the angels and are the children of God, being the children of the resurrection.

12. Now that the dead are raised even Moses shewed at the bush, when he called the Lord, the God Abraham, Isaac and Jacob, for he is not the God of the dead, but of the living, for all live unto Him.

LECTION 53. 3. -The healing of the blind by means of clay mingled with saliva is mentioned by ancient physicians. Vespasian is said to have cured by this means. This shows that Jesus did not hesitate to employ natural remedies, when they were likely to effect their purpose.

Lection 54

The Examination of Him Who was Born Blind.

1. THEN they brought to the Pharisees him that aforetime was blind. And it was the Sabbath day when Iesus made the clay, and opened his eyes.

2. Then again the Pharisees also asked him how he had received his sight. He said unto them, He put clay upon mine eyes, and I washed, and do see.

3. Therefore said some of the Pharisees, This man is not of God, because he keepeth not the Sabbath day. Others said, how can a man that is a sinner do such miracles? And there was a division among them.

4. They say unto the blind man again, What sayest thou of him, that he hath opened thine eyes? He said, He is a prophet.

5. But the Jews did not believe concerning him, that he had been blind, and received his sight, until they called the parents of him that had received his sight.

6. And they asked them, saying, Is this your son, who ye say was born blind? how then doth he now see? His parents answered them and said, We know that this is our son, and that he was born blind; but by what means he now seeth we know not; nor who hath opened his eyes; he is of age; ask him, he shall speak for himself.

7. These words spake his parents, because they feared the Jews; for the Jews had agreed already, that if any man did confess that he was the Christ he should be put out of the synagogue. Therefore said his parents, He is of age? ask him.

8. Then again called they the man that was blind, and said unto him, Give God the praise: we know that this man is a sinner. He answered and said, Whether he be a sinner or no, I know not; one thing I know, that, whereas I was blind, now I see.

9. Then said they to him again, What did he to thee? how opened he thine eyes? He answered them, I have told you already, and ye did not hear: wherefore would ye hear it again? will ye also be his disciples?

10. Then they reviled him, and said, Thou art his disciple; but we are Moses' disciples. We know that God spake unto Moses: as for this fellow, we know not from whence he is.

11. The man answered and said unto them, Why herein is a marvellous thing, that ye know not from whence he is, and yet he hath opened mine eyes. Now we know that God heareth not sinners;

12. But if any man be a worshipper of God, and doeth his will, him he heareth. Since the world began was it not heard that any man opened the eyes of one that was born blind. If this man were not of God, he could do nothing.

13. They answered and said unto him, Thou wast altogether born in sins, and dost thou teach us? And they cast him out.

14. Iesus heard that they had cast him out; and when he had found him, he said unto him, Dost thou believe on the Son of God ? He answered and said, Who is he, Lord, that I might believe on him.

15. And Iesus said unto him, Thou hast both seen him, and it is he that talketh with thee. And he said, Lord, I believe. And he worshipped him.

16. And Iesus said, For judgment I am come into this world, that they which see not might see; and that they which see might be made blind. And some of the Pharisees which were with him heard these words, and said unto him, Are we blind also?

17. AND Iesus, when he came to a certain place where seven palm trees grew, gathered his disciples around him, and to each he gave a number and a name which he only knew who received it. And he said unto them, Stand ye as pillars in the House of God, and shew forth the order according to your numbers which ye have received.

18. And they stood around him, and they made a body four square, and they counted the number, and could not. And they said unto him, Lord we cannot. And Jesus said, Let him who is greatest among you be even as the least, and the symbol of that which is first be as the symbol of that which is last.

19. And they did so, and in every way was there equality, and yet each bore a different number and the one side was as the other and the upper was as the lower, and the inner as the outer. And the Lord said, It is enough. Such is the House of the wise Master Builder. Foursquare it is, and perfect. Many are the Chambers, but the House is One.

20. Again consider the Body of man, which is a Temple of the Spirit. For the body is one, united to its head, which with it is one body. And it has many members, yet, all are one body and the one Spirit ruleth and worketh in all; so also in the kingdom.

21. And the head doth not say to the bosom, I have no need of thee, nor the right hand to the left, I have no need of thee, nor the left foot to the right, I have no need of thee; neither the eyes to the ears, we have no need of you, nor the mouth to the nose, I have no need for thee. For God hath set in the one body every member as is fitting.

22. If the whole were the head, where were the breasts? If the whole were the belly, where were the feet? yea, those members which some affirm are less honourable, upon them hath God bestowed the more honour.

23. And those parts which some call uncomely, upon them hath been bestowed more abundant comeliness, that they may care one for the other; so, if one member suffers, all members suffer with it, and if one member is honoured all members rejoice.

24. Now ye are my Body; and each one of you is a member in particular, and to each one of you do I give the fitting place, and one Head over all, and one Heart the centre of all, that there be no lack nor

schism, that so with your bodies, your souls and your spirits ye may glorify the All Parent through the Divine Spirit which worketh in all and through all.

LECTION 54. 1-13. -The wrangling of the Pharisees over this case of healing has its parallels in our times in the Churches which assign to the devil all that they cannot comprehend, and cut out the Healer as a sinner and a heretic, denying the power of God in Man.

LECTION 54. 14. -This is one of those "parables and dark sayings" of him who spake as never man spake. The words taken literally suggest to the mind a perfect crystal sphere, and by correspondence, a perfect man or woman- in modern phrase "an all rounder," one who views things not from one side only, but from every side. There are many who keep the law in one or more points, but neglect all the rest; or keep it in all points but the one which is against their own particular failing -who "compound for sins they are inclined to, by damming those they have no mind to." **Cont.**

LECTION 54. 17-20. -The meaning of these words and this action is very obscure, but if we describe the magic square of 7, it seems to make it intelligible as the mystic symbol of him who regarded everything by number and by measure, and which seems to have reference to the period of his mortal life, 49 years, as well as the number of the Council, Cardinals and Priests of the Church universal, 48, presided over by its Head, 49, which the action of Iesus seemed to symbolize, and in a way, foreshadow. *See square*

LECTION 54. 21-24. -Here we have the original words of Christ, from which Paul adopted his simile in Rom. xii., and In 1 Cor. xii.

Lection 55

Christ The Good Shepherd. One With the Father.

1. AT that time there passed by the way a shepherd leading his flock to the fold; and Iesus took up one of the young lambs in his arms and talked to it lovingly and pressed it to his bosom. And he spake to his disciples saying:

2. I am the good shepherd and know my sheep and am known of mine. As the Parent of all knoweth me, even so know I my sheep, and lay down my life for the sheep. And other sheep I have, which are not of this fold; them also must I bring, and they shall hear my voice, and there shall be one flock and one shepherd.

3. I lay down my life, that I may take it again. No man taketh it from me, but I lay it down of myself. I have power to lay my body down and I have power to take it up again.

4. I am the good shepherd; the good shepherd feedeth his flock, he gathereth his lambs in his arms and carrieth them in his bosom and gently leadeth those that are with young, yea the good shepherd giveth his life for the sheep.

5. But he that is an hireling, and not the shepherd, whose own the sheep are not, seeth the wolf coming and leaveth the sheep and fleeth, and the wolf catcheth them and scattereth the sheep. The hireling fleeth because he is an hireling and careth not for the sheep.

6. I am the door: by me all who enter shall be safe, and shall go in and out and find pasture. The evil one cometh not but for to steal and to kill and destroy; I am come that they might have life, and that they might have it more abundantly.

7. He that entereth in by the door, is the shepherd of the sheep, to whom the porter openeth, and the sheep hear his voice, and he calleth his sheep by name, and leadeth them out, and he knoweth the number.

8. And when he putteth forth his sheep he goeth before them and the sheep follow him for they know his voice. And a stranger will they not follow, but will flee from him, for they know not the voice of strangers.

9. This parable spake Iesus unto them, but they understood not what things they were which he spake unto them. Then said Iesus unto them again, My sheep hear my voice, and I know them, and they follow me and I give unto them eternal life and they shall never perish, neither shall any man pluck them out of

my hand.

10. My Parent who gave them me, is greater than all and no man is able to pluck them out of my Parent's hand. I and my Parent are One.

11. Then the Jews took up stones again to stone him. Jesus answered them, Many good works have I shewed you from my Parent, for which of those works do ye stone me?

12. The Jews answered him, saying, For a good work we stone thee not, but for blasphemy, because that thou being a man maketh thyself equal with God. Jesus answered them, Said I that I was equal to God? nay, but I am one with God. Is it not written in the Scripture, I said, Ye are gods?

13. If he called them gods, unto whom the word of God came, and the Scripture cannot be broken, say ye of him, whom the Parent of all hath sanctified and sent into the world. Thou blasphemest; because I said I am the Son of God, and therefore One with the All Parent?

14. If I do not the works of my Parent believe me not, but if I do, though ye believe not me, believe the works, that ye may know and believe that the Spirit of the great Parent is in me, and I in my Parent.

15. Therefore they sought again to take him, but he escaped out of their hands and went away again beyond Jordan, into the place where John at first baptized and there he abode.

16. And many resorted unto him, and said, John, indeed did not miracle, He is the Prophet that should come. And many believed on him.

LECTION 55. 1. -This beautiful parable bas been sadly mangled in the A.V., and shorn of the opening incident which led to the discourse.

Lection 56

The Raising of Lazarus.

1. Now a certain man was sick, named Lazarus of Bethany, the town of Mary and her sister Martha. (It was that Mary who anointed the Lord with ointment and wiped his feet with her hair, whose brother Lazarus was sick).

2. Therefore his sisters sent unto him saying, Lord, behold he whom thou lovest is sick. When Iesus heard that, he said, This sickness is not unto death, but that the glory of God might be manifest in him. Now Jesus loved Mary and her sister and Lazarus.

3. When he heard that he was sick, he abode two days still in the same place where he was. Then after that, saith he to his disciples, Let us go into Judea again.

4. His disciples said unto him, Master, the Jews of late sought to stone thee and goest thou thither again? Iesus answered, Are there not twelve hours in the day? If any man walketh in the day he stumbleth not, because he seeth the light of this world.

5. But if a man walk in the night, he stumbleth, because there is no light in him. These things said he, and after that he saith unto them, Our friend Lazarus sleepeth, but I go that I may awake him out of sleep.

6. Then said his disciples, Lord if he sleep, he shall do well. And a messenger came unto him saying, Lazarus is dead.

7. Now when Iesus came, he found that he had lain in the grave four days already (Bethany was nigh unto Jerusalem, about fifteen furlongs off). And many of the Jews came to Martha and Mary to comfort them concerning their brother.

8. Then Martha, as soon as she heard that Iesus was coming, went and met him, but Mary sat still in the house. Then said Martha unto Iesus, Lord if thou hadst been here my brother had not died. But I know that even now, whatsoever thou wilt ask of God, God will give it thee.

9. Iesus saith unto her, Thy brother sleepeth, and he shall rise again. Martha said unto him, I know that he shall rise again, at the resurrection at the last day.

10. Iesus said unto her, I am the resurrection and the life, he that believeth in me, though he were dead yet shall he live. I am the Way, the Truth and the Life, and whosoever liveth and believeth in me shall never die.

11. She saith unto him, Yea, Lord : I believe that thou art the Christ, the Son of God, which should come

into the world. And when she had so said she went her way and called Mary her sister secretly saying, The Master is come and calleth for thee. As soon as she heard that she arose quickly and came unto him.

12. Now Iesus was not yet come into the town, but was in that place where Martha met him. The Jews then which were with her in the house and comforted her, when they saw Mary that she arose up hastily and went out, followed her saying, She goeth unto the grave to weep there.

13. Then when Mary was come to where Iesus was, and saw him she fell down at his feet, saying unto him, Lord if thou hadst been here my brother had not died. When Jesus therefore saw her weeping and the Jews also weeping that came with her, he groaned in the spirit and was troubled. And said, Where have ye laid him? They said unto him, Lord, come and see, and Jesus wept.

14. Then said the Jews, Behold, how he loved him! And some of them said, Could not this man which opened the eyes of the blind, have caused that even this man should not have died? Iesus therefore groaning again in himself (for he feared that he might be already dead) cometh to the grave. It was a cave and a stone lay upon it.

15. Iesus said, Take ye away the stone. Martha, the sister of him supposed to be dead, saith unto him, Lord by this time he stinketh, for he hath been dead four days. Iesus saith unto her, Said I not unto thee, that if thou wouldest believe thou shouldst see the glory of God? Then they took away the stone from the place where Lazarus was laid.

16. And Iesus lifted up his eyes and chanting, invoked the great Name, and said, My Parent, I thank Thee that thou has heard me. And I know that Thou hearest me always, but because of the people which stand by I call upon Thee that they may believe that Thou hast sent me. And when he had thus spoken he cried with a loud voice, Lazarus come forth.

17. And he that was as dead came forth bound hand and foot with graveclothes, and his face was: bound about with a napkin.

18. Iesus said unto them, Loose him and let him go. When the thread of life is cut indeed, it cometh not again, but when it is whole there is hope. Then many of the Jews which came to Mary and had seen the things which Iesus did, believed on him.

LECTION 56. -This touching account of the raising of Lazarus is here given as it took place. The verses 13-16 in the Authorized Version are an evident interpolation to magnify the occasion, for, being omitted, the narrative is unbroken and complete without them. As with the daughter of Lazarus, so with Lazarus, he was carried to his burial in a state of trance, indistinguishable from death, and by his friends believed to be dead. At the present time in countries where there are mortuaries or waiting rooms for the dead, it is found that five per thousand recover on their way to burial who otherwise would have been buried alive.

Lection 57

Concerning little Children.
The Forgiveness of Those Who Trespass.
Parable of the Fishes.

1. AT the same time came the disciples unto Iesus, saying, who is the greatest in the kingdom of Heaven? And Iesus called a little child unto him and set him in the midst of them and said, Verily I say unto you, except ye be converted and become innocent and teachable as little children, ye shall not enter into the kingdom of Heaven.

2. Whosoever therefore shall humble himself as this little child, the same is the greatest in the kingdom of Heaven. And whoso shall receive one such little child in my name receiveth me.

3. Woe unto the world because of offenses! for it must needs be that offences come, but woe to that man by whom the offence cometh. Wherefore if thy lust, or thy pleasure do offend others, cut them off and cast them from thee, it is better for thee to enter into life without, rather than having that which will be cast into everlasting fire.

4. Take heed that ye neglect not one of these little ones, for I say unto you, That in heaven their angels

do always behold the Face of God. For the Son of man is come to save that which was lost.

5. How think ye? if a man have a hundred sheep, and one of them be gone astray, doth he not leave the ninety and nine and go into the mountains and seek that which is gone astray? And if so be that he find it, verily I say unto you, he rejoiceth more over that sheep than over the ninety and nine which went not astray.

6. Even so it is not the will of your Parent, Who is in heaven, that one of these little one should perish.

7. AND there were certain men of doubtful mind, came unto Iesus, and said unto him: Thou tellest us that our life and being is from God, but we have never seen God, nor do we know of any God. Canst thou shew us Whom thou callest the Father-Mother, one God? We know not if there be a God.

8. Iesus answered them, saying, Hear ye this parable of the fishes. The fishes of a certain river communed with one another, saying, They tell us that our life and being is from water, but we have never seen water, we know not what water is. Then some among them, wiser than the rest, said: We have heard there dwelleth in the sea a wise and learned Fish, who knoweth all things. Let us journey to him, and ask him to shew us what water is.

9. So several of them set out to find this great and wise Fish and they came at last to the sea wherein the wise Fish dwelt, and they asked of him.

10. And when he heard them he said unto them, O ye foolish fish that consider not! Wise are ye, the few, who seek. In the water ye live, and move, and have your being; from the water ye came, to the water ye return. Ye live in the water, yet ye know it not. In like manner, ye live in God, and yet ye ask of me, "Shew us God." God is in all things, and all things are in God.

11. AGAIN Iesus said unto them, If thy brother or sister shall trespass against thee, go and declare the fault between thee and thy brother or sister alone; if they shall hear thee, thou hast gained them. But if they will not hear thee, then take with thee one or two more, that in the mouth of two or three witnesses every word may be established.

12. And if they shall neglect to hear them, tell it unto the church, but if they neglect to hear them, tell it unto the church, but if they neglect to hear the church, let them be unto thee as those that are outside the church. Verily I say unto you, Whatsoever ye shall justly bind on earth, shall be bound in heaven, and whatsoever ye shall justly loose in earth, shall be loosed in heaven.

13. Again I say unto you, That if seven, or even if three of you shall agree on earth as touching anything that they ask, it shall be done for them of my Father-Mother Who is in heaven. For where even three are gathered together in my name there I am in the midst of them, and if there be but one, I am in the heart of that one.

14. THEN came Peter to him and said, Lord, how oft shalt my brother sin against me and I forgive him? till seven times? Iesus saith unto him, I say not unto thee, Until seven times, but until seventy times seven. For in the Prophets likewise unrighteousness was found, even after they were anointed by the Holy Spirit.

15. And he spake this parable, saying, There was a certain king who would take account of his servants, and when he had begun to reckon, one was brought unto him which owed him ten thousand talents. But forasmuch as he had not to pay, his lord commanded him to be sold, and his wife and children and all that he had, and payment to be made.

16. The servant therefore, fell down and worshipped him, saying, Lord, have patience with me and I will pay thee all. Then the lord of that servant was moved with compassion and loosed him, and forgave him his debt.

17. But the same servant went out and found one of his fellow-servants which owed him a hundred pence, and he laid hands on him and took him by the throat, saying, Pay me that thou owest.

18. And his fellow-servant fell down at his feet and besought him, saying, Have patience with me and I will pay thee all. And he would not, but went and cast him into prison till he should pay the debt.

19. So when his fellow-servants saw what he had done they were very sorry, and came and told unto their lord all that was done.

20. Then his lord, after he had called him, said unto him, O thou wicked servant, I forgave thee all that debt because thou desiredst me; shouldst not thou also have had compassion on thy fellow-servant, even as I had pity on thee. And his lord was wroth, and delivered him to the tormentors, till he should pay all that was due unto him.

21. So likewise shall the heavenly Parent judge you, if ye from your hearts forgive not every one, his brother or sister, their trespasses. Nevertheless, let every man see that he pay that which he oweth, for God loveth the just.

LECTION 57. 4. -The doctrine of guardian angels receives full support from these words. But the Churches of the so-called Reformation have flung away this consoling and helpful belief, with other doctrines of the Christian Church in all ages, the truth of which science and occultism are now showing.

Lection 58

Divine Love To The Repentant.

1. Iesus said unto the disciples and to the multitude around them, Who is the son of God? Who is the daughter of God? Even the company of them who turn from all evil and do righteousness, love mercy and walk reverently with their God. These are the sons and the daughters of man who come up out of Egypt, to whom it is given that they should be called the sons and the daughters of God.

2. And they are gathered from all tribes and nations and peoples and tongues, and they come from the East and the West and the North and the South, and they dwell on Mount Zion, and they eat bread and they drink of the fruit of the vine at the table of God, and they see God face to face.

3. Then drew near unto him all the taxgatherers and sinners for to hear him. And the Pharisees and Scribes murmured, saying, This man receiveth sinners and eateth with them.

4. AND he spake this parable unto them, saying, What man of you having an hundred sheep, if he lose one of them doth not leave the ninety and nine in the wilderness, and go after that which is lost, until he find it? And when he hath found it he layeth it on his shoulders, rejoicing.

5. And when he cometh home, he calleth together his friends and neighbours, saying unto them, Rejoice with me, for I have found my sheep which was lost. I say unto you, that likewise joy shall be in heaven over one sinner that repenteth, more than over ninety and nine just persons which need no repentance.

6. Either what woman having ten pieces of silver, if she lose one piece doth not light a candle and seek diligently till she find it? And when she hath found it she calleth her friends and her neighbours together, saying, Rejoice with me, for I have found the piece of silver which I had lost. Likewise, I say unto you, there is joy in the presence of the angels of God over one sinner that repenteth.

7. AND he also spake this parable, A certain man had two sons, and the younger of them said to his parents, Give me the portion of goods that falleth to me. And they divided unto him their living. And not many days after the younger son gathered all together and took his journey into a fair country, and there wasted his substance with riotous living.

8. And when he had spent all, there arose a mighty famine in that land, and he began to be in want. And he went and joined himself to a citizen of that country, and he sent him into his fields to feed swine. And he would fain have filled his body with the husks that the swine did eat, and no man gave unto him.

9. And when he came to himself he said, How many hired servants of my father's have bread enough and to spare, and I perish with hunger! I will arise and go to my father and mother, and will say unto them. My father and my mother, I have sinned against Heaven and before you, and am no more worthy to be called your son, make me as one of your hired servants.

10. And he arose and came to his parents. But when he was a great way off, his mother and his father saw him and had compassion, and ran and fell on his neck and kissed him. And the son said unto them, My father and my mother, I have sinned against Heaven and in your sight, and am no more worthy to be called your son.

11. But the father said to his servants, Bring forth the best robe, and put it on him, and put a ring on his hand and shoes on his feet, and bring hither the best ripe fruits, and the bread and the oil and the wine, and let us eat and be merry; for this my son was dead and is alive again, he was lost and is found. And they began to be merry.

12. Now his elder son was in the field, and as he came and drew nigh to the house he heard music and dancing. And he called one of the servants and asked what these things meant. And he said unto him, Thy brother who was lost is come back, and thy father and thy mother have prepared the bread and the oil and the wine and the best ripe fruits, because they have received him safe and sound.

13. And he was angry and would not go in, therefore came his father out and entreated him. And he answering, said to his father, Lo, these many years have I served thee, neither transgressed I at any time

thy commandments, and yet thou never gavest me such goodly feast that I may make merry with my friends.

14. But as soon as this thy son is come, which hath devoured thy living with harlots, thou preparest for him a feast of the best that thou hast.

15. And his father said unto him, Son, thou art ever with me, and all that I have is thine. It was meet, therefore, that we should be merry and be glad, for this thy brother was dead and is alive again, and was lost and is found.

LECTION 58. 2. -The charity and comprehensiveness of the true doctrine of Jesus here manifests themselves. It is not a mere narrow creed or belief, but true repentance which merits the forgiveness of God.

Lection 59

Iesus Forewarneth His Disciples. He Findeth Zaccheus.

1. AND Iesus went up into a mountain and there he sat with his disciples and taught them, and he said unto them, Fear not, little flock, for it is your Father's good pleasure to give you the kingdom.

2. Sell that ye have and do that which is good, for them which have not; provide yourselves bags which wax not old, a treasure in the heavens that faileth not, where no thief approacheth, neither moth corrupteth. For where your treasure is, there will your heart be also.

3. Let your loins be girded about, and your lights burning, and ye yourselves like unto men that wait for their lord, when he will return from the wedding that when he cometh and knocketh they may open unto him immediately.

4. Blessed are those servants whom the lord, when he cometh, shall find watching; verily I say unto you that he shall gird himself and make them to sit down at his table, and will come forth and serve them.

5. And if he shall come in the second watch, or come in the third watch and find them so, blessed rare those servants.

6. And this know, that the guardian of the house not knowing what hour the thief would come, would have watched and not have suffered his house to have been broken through. Be ye therefore ready also, for the Son of man cometh at an hour when ye think not.

7. Then Peter said unto him, Lord, speakest thou this parable unto us, or even to all? And the Lord said, Who then is that faithful and wise steward, whom his lord shall make ruler over his household, to give them who serve their portion in due season?

8. Blessed is that servant whom his lord when he cometh shall find so doing. Of a truth I say unto you, that he will make him ruler over all that he hath.

9. But and if that servant say in his heart, My lord delayeth his coming and shall begin to beat the menservants and maidservants and to eat and drink and to be drunken, the lord of that servant will come in a day when he looketh not for him, and at an hour when he is not aware and will appoint him his portion with the unfaithful.

10. And that servant which knew his lord's will and prepared not himself, neither did according to his will, shall be beaten with many stripes. But he that knew not, and did commit things worthy of stripes, shall be beaten with few stripes. For unto whomsoever much is given, of him shall they much require the less.

11. For they who know the Godhead, and have found in the way of Life the mysteries of light and then have fallen into sin, shall be punished with greater chastisements than they who have not known the way of Life.

12. Such shall return when their cycle is completed and to them will be given space to consider, and amend their lives, and learning the mysteries, enter into the kingdom of light.

13. AND Iesus entered and passed through Jericho. And, behold, there was a man named Zaccheus, which was the chief among the collectors of tribute, and he was rich.

14. And he sought to see Iesus who he was; and could not for the press, because he was little of stature. And he ran before, and climbed up into a sycamore tree to see him: for he was to pass that way.

15. And when Iesus came to the place, he looked up, and saw him, and said unto him, Zacheus, make haste, and come down; for to day I must abide at thy house. And he made haste and came down, and received him joyfully.

16. And when they saw it, they all murmured, saying, That he was gone to be guest with a man that is a sinner.

17. And Zachaeus stood, and said unto the Lord, Behold, Lord, the half of my goods I give to the poor; and if I have taken anything from any man by false accusation, I restore him fourfold.

18. And Iesus said unto him, This day is salvation come to thine house, forsomuch as thou art a just man, thou also art a son of Abraham. For the Son of man is come to seek and to save that which ye deem to be lost.

LECTION 59. 1l-12. -The teaching of our Lord as to cycles, and the unity of life, in many existences, has been suppressed for long ages, but now sees the light, at the end of the cycle.

Lection 60

Iesus Rebuketh Hypocrisy.

1. THEN spake Iesus to the multitude, and to his disciples, saying. The scribes and the Pharisees sit in Moses's seat. All therefore whatsoever they bid you observe, that observe and do; but do not ye after their works: for they say and do not. For they bind heavy burdens and grievous to be borne, and lay them on men's shoulders; but they themselves will not move them with one of their fingers.

2. But all their works they do for to be seen of men; they make broad their phylacteries, and enlarge the borders of their garments, and love the uppermost rooms at feasts, and the chief seats in the synagogues, and greetings in the markets, and to be called of men, Rabbi, Rabbi.

3. But desire not ye to be called Rabbi: for one is your Rabbi, even Christ; and all ye are brethren. And call not any one father on earth, for on earth are fathers in the flesh only; but in Heaven there is One Who is your Father and your Mother, Who hath the Spirit of truth, Whom the world cannot receive.

4. Neither desire ye to be called masters, for one is your Master, even Christ. But they that are greatest among you shall be your servants. And whosoever shall exalt themselves shall be abased; and they that are humble in themselves shall be exalted.

5. Woe unto you, scribes and Pharisees, hypocrites! for ye shut up the kingdom of Heaven against men: for ye neither go in yourselves neither suffer ye them that are entering, to go in,

6. Woe unto you, scribes and Pharisees, hypocrites" for ye devour widows' houses, and for a pretence make long prayer; therefore ye shall receive the greater damnation.

7. Woe unto you, scribes and Pharisees, hypocrites! for ye compass sea and land to make one proselyte; and when he is made, ye make him twofold more the child of hell than yourselves.

8. Woe unto you, ye blind guides, who say, Whosoever shall swear by the Temple, it is nothing, but whosoever shall swear by the gold of the Temple, he is a debtor! Ye fools and blind; for whether is greater, the gold, or the Temple that sanctifieth the gold?

9. And, Whosoever shall swear by the altar, it is nothing; but whosoever sweareth by the gift that is upon it, he is guilty. Ye fools and blind: for whether is greater, the gift, or the altar, that sanctifieth the gift?

10. Whoso therefore shall swear by the altar, sweareth by it, and by all things thereon. And whoso shall swear by the Temple, sweareth by it, and by him that dwelleth therein. And he that shall swear by Heaven sweareth by the throne of God, and by the Holy One that sitteth thereon.

11. Woe unto you, scribes and Pharisees, hypocrites! for ye pay tithe of mint and anise and cummin, and have omitted the weightier matters of the law, judgment, mercy, and faith: these ought ye to have done, and not to leave the other undone. Ye blind guides! for ye strain out a gnat, and swallow a camel.

12. Woe unto you, scribes and Pharisees, hypocrites! for ye make clean the outside of the cup and of the platter, but within they are full of extortion and excess. Thou blind Pharisee, cleanse first that which is

within the cup and platter, then the outside of them that they may be clean also.

13. Woe unto you, scribes and Pharisees, hypocrites! for ye are like unto whited sepulchres, which indeed appear beautiful outward, but are within full of the bones of the dead and of all uncleanness. Even so ye also outwardly appear righteous unto men, but within ye are full of hypocrisy and make believe.

14. Woe unto you, scribes and Pharisees, hypocrites! because ye build the tombs of the prophets, and garnish the sepulchres of the righteous, and say, If we had been in the days of our fathers, we would not have been partakers with them in the blood of the prophets.

15. Wherefore ye be witness unto yourselves, that ye do as the children of them which killed the prophets. Fill ye up then the measure of your fathers.

16. Wherefore saith holy Wisdom, behold I send unto you prophets, and wise men, and scribes: and some of them ye shall kill and crucify; and some of them shall ye scourge in your synagogues, and persecute them from city to city. And upon you shall come all the righteous blood shed upon the earth, from the blood of righteous Abel unto the blood of Zacharias son of Barachias, who was slain between the temple and the altar. Verily I say unto you, All these things shall come upon this generation.

17. O Jerusalem, Jerusalem, thou that killest the prophets, and stonest them which are sent unto thee, how often would I have gathered thy children together, even as a hen gathereth her chickens under her wings, and ye would not!

18. Behold, now your house is left unto you desolate. For I say unto you, Ye shall not see me henceforth, till ye shall say, Holy, Holy, Holy, Blessed are they who come in the Name of the Just One.

LECTION 60. 16. -The same Zaccharias who is mentioned in the beginning as the father of John the Baptist (see Note 111-2), also the Proto Evangelism attributed to James, the Bishop of Jerusalem.

The Gospel of the Holy Twelve
Translated from the original Aramaic
by Rev. G.J.R. Ouseley

Section 7, Lections 61 thru 70

Lection 61

Jesus Foretells The End Of The Cycle.

1. AND as Jesus sat upon the Mount of Olives, the disciples came unto him privately, saying, Tell us, when shall these things be? and what shall be the sign of thy coming, and of the end of the world? And Jesus answered and said unto them, Take heed that no man deceive you. For many shall come in my Name, saying, I am Christ; and shall deceive many.

2. And ye shall hear of wars and rumours of wars; see that ye be not troubled; for all these things must come to pass, but the end is not yet. For nation shall rise against nation, and kingdom against kingdom; and there shall be famines, and pestilences, and earthquakes, in divers places. All these are the beginning of sorrows.

3. And in those days those that have power shall gather to themselves the lands and riches of the earth for their own lusts, and shall oppress the many who lack and hold them in bondage, and use them to increase their riches, and they shall oppress even the beasts of the field, setting up the abominable thing. But God shall send them his messenger and they shall proclaim his laws, which men have hidden by their traditions, and those that transgress shall die.

4. Then shall they deliver you up to be afflicted, and shall kill you; and ye shall be hated of all nations for my Name's sake. And then shall many be offended, and shall betray one another, and shall hate one another. And many false prophets shall rise, and shall deceive many.

5. And because iniquity shall abound, the love of many shall wax cold. But he that shall endure unto the end, the same shall be saved. And this gospel of the kingdom shall be preached in all the world for a witness unto all nations; and then shall the end come.

6. When ye therefore shall see the abomination of desolation, spoken of by Daniel the prophet, stand in the holy place, (whoso readeth, let him understand) then let them which be in Judea flee to the mountains. Let them which are on the housetop not come down to take anything out of the house; neither let them who are in the field return back to take their clothes.

7. And woe unto them that are with child, and to them that give suck in those days! But pray ye that your fight be not in the winter, neither on the Sabbath day; for there shall be great tribulation, such as was not since the beginning of the world to this time, no, nor ever shall be. And except those days be shortened, there should no flesh be saved; but for the elect's sake those days shall be shortened.

8. Then if any man shall say unto you, Lo, here is Christ, or there; haste not to believe. For there shall arise false Christs, and false prophets, who shall shew great signs and wonders; insomuch that, if it were possible, they shall deceive the very elect. Behold, I have told you before.

9. Wherefore if they shall say unto you, Behold, he is in the desert; go not forth: behold, he is in the secret chambers; haste not to believe. For as the lightening cometh out of the east, and shineth even unto the west; so shall also the coming of the Son of man be. For wheresoever the carcass is, there will the eagles be gathered together.

10. Immediately after the tribulation of those days shall the sun be darkened, and the moon shall not give her light, and the stars shall fall from Heaven, and the powers of the Heavens shall be shaken.

11. And then shall appear the sign of the Son of man in Heaven; and then shall all the tribes of the earth mourn, and they shall see the Son of man coming in the clouds of Heaven with power and great glory. And he shall send his angels with a great sound as of a trumpet, and they shall gather together his elect from the four winds, from one end of Heaven to the other.

12. Now learn a parable of the fig tree; When its branch is yet tender, and putteth forth leaves, ye know that summer is nigh. So likewise ye, when ye shall see all these things, know that it is near, even at the doors. Verily I say unto you, this generation shall not pass till all these things be fulfilled. Heaven and earth shall pass away, but my words shall not pass away.

13. But of that day and hour knoweth no man, no, not the angels of Heaven, but the All Parent only. For as the days of Noe were, so shall also the coming of the Son of man be.

14. For as in the days that were before the flood, they were eating and drinking, marrying and giving in marriage, until the day that Noe entered into the ark and knew not until the flood came, and took them all away; so shall also the coming of the Son of man be.

15. Then shall two be in the field; the one shall be taken, and the other left. Two women shall be grinding at the mill; the one shall be taken, and the other left. Watch therefore: for ye know not what hour your Lord doth come.

16. But know this, that if the guardian of the house had known in what watch the thief would come, he would have watched, and would not have suffered his house to be broken up. Therefore be ye also ready: for in such an hour as ye think not, the Son of man cometh.

17. Who then is a faithful and wise servant, whom his lord hath made ruler over his household, to give them meat in due season?

18. Blessed be that servant, whom his lord when he cometh shall find so doing. Verily I say unto you, That he shall make him ruler over all his goods.

19. But and if that evil servant shall say in his heart, My lord delayeth his coming, and shall begin to smite his fellow servants, and to eat with the glutton, and drink with the drunken.

20. The lord of that servant shall come in a day when he looketh not for him, and in an hour that he is not aware of. And shall appoint him his portion with the hypocrites in the outer darkness with the cruel, and them that have no love, no pity: there shall be weeping and gnashing of teeth.

LECTION 61. -All through this chapter the language is highly symbolical, but will present little difficulty to the initiated. *v.* 12. (*See* Note in the original "Genesis" Edited by same Editor).

Lection 62

The Parable Of The Ten Virgins.

1. THEN shall the kingdom of Heaven be like unto ten virgins, which took their lamps, and went forth to meet the bridegroom. And five of them were wise, and five were foolish.

2. They that were foolish took their lamps, and took no oil with them: But the wise took oil in their vessels with their lamps.

3. While the bridegroom tarried, they all slumbered and slept. And at midnight there was a great cry made, Behold, the bridegroom cometh; go ye out to meet him. Then all those virgins arose, and trimmed their lamps.

4. And the foolish said unto the wise, Give us of your oil; for our lamps are gone out. But the wise answered, saying, Not so, lest there be not enough for us and you: but go ye rather to them that sell, and buy for yourselves.

5. And while they went to buy, the bridegroom came; and they that were ready went in with him to the marriage: and the door was shut.

6. Afterwards came also the other virgins, saying Lord, Lord, open to us. But he answered and said, Verily I say unto you. I know you not.

7. Watch therefore, for ye know neither the day nor the hour wherein the Son of man cometh. Keep your lamps burning.

LECTION 62. 1. -This parable of the ten virgins most accurately indicates the oblivion and indifference which shall come on Christians in the last days of the Christian Church -the days of Laodicean indifference, the Seventh or last age.

Lection 63

Parable Of The Talents

1. He also said: The kingdom of Heaven is as a man traveling into a far country, who called his own servants, and delivered unto them his goods. And unto one he gave five talents, to another two, and to another one; to every man according to his several ability; and straightway took his journey.

2. Then he that had received the five talents went and traded with the same, and made them other five talents. And likewise he that had received two, he also gained other two. But he that had received one went and digged in the earth, and hid his lord's money.

3. After a long time, the lord of those servants cometh, and reckoneth with them. And so he that had received five talents came and brought other five talents, saying, Lord, thou deliveredst unto me five talents; behold, I have gained beside them five talents more. His lord said unto him, Well done, thou good and faithful servant: thou hast been faithful over a few things, I will make thee ruler over many things; enter thou into the joy of thy lord.

4. He also that had received two talents came and said, Lord, thou deliveredst unto me two talents; behold, I have gained two other talents beside them. His lord said unto him, Well done, good and faithful servant; thou hast been faithful over a few things, will make thee ruler over many things; enter thou into the joy of thy lord.

5. Then he which had received the one talent came and said, Lord, I knew thee that thou art an hard man, reaping where thou hast not sown, and gathering where thou hast not strawed. And I was afraid, and went and hid thy talent in the earth; lo, there thou hast that is thine.

6. His lord answered and said unto him, Thou wicked and slothful servant, dost thou tell me that I reap where I sowed not, and gather where I have not strawed? Thou oughtest therefore to have put thy talents to use, with profit, and then at my coming I should have received mine own with usury.

7. Take therefore the talent from him, and give it unto him who hath two talents. For unto every one that hath improved shall be given, and he shall have abundance, but from him that hath not improved, shall be taken away, even that which he hath. And cast yet out the unprofitable servant into outer darkness, for that is the portion he hath chosen.

8. Jesus also said unto his disciples, Be ye approved money-changers of the kingdom, rejecting the bad and the false, and retaining the good and the true.

9. AND Jesus sat over against the Treasury and beheld how the people cast money into the Treasury.

10. And there came a certain poor widow and she threw in two mites, which make a farthing.

11. And He called His disciples unto him and said, Verily I say unto you, that this poor widow hath cast more in than all they which have cast into the Treasure.

12. For all they did cast in of their abundance, but she of her poverty did cast in all that she had, even all her living.

LECTION 63. 8. -These words are one of the "last sayings" of Jesus and vividly describe the duty of a Christian Council, so oft neglected.

Widow's Mites; tiny copper coins circulated in Judaea during the time of Christ.

The Gospel Of The Holy 12

Lection 64

Jesus Teacheth In The Palm Circle.

1. JESUS came to a certain fountain near Bethany, around which grew twelve palm trees, where he often went with his disciples to teach them of the mysteries of the kingdom, and there he sat beneath the shade of the trees and his disciples with him.

2. And one of them said, Master, it is written of old, The Alohim made man in Their own image, male and female created They them. How sayest thou then that God is one? And Iesus said unto them, Verily, I said unto you, In God there is neither male nor female and yet both are one, and God is the Two in One. He is She and She is He. The Alohim—our God—is perfect, Infinite, and One.

3. As in the man, the Father is manifest, and the Mother hidden; so in the woman, the Mother is manifest, and the Father hidden. Therefore shall the name of the Father and the Mother be equally hallowed, for They are the great Powers of God, and the one is not without the other, in the One God.

4. Adore ye God, above you, beneath you, on the right hand, on the left hand before you, behind you, within you, around you. Verily, there is but One God, Who is All in All, and in Whom all things do consist, the Fount of all Life and all Substance, without beginning and without end.

5. The things which are seen and pass away are The manifestations of the unseen which are eternal, that from the visible things of Nature ye may reach to the invisible things of the Godhead; and by that which is natural, attain to that which is spiritual.

6. Verily, the Alohim created man in the divine image male and female, and all nature is in the Image of God, therefore is God both male and female, not divided, but the Two in One, Undivided and Eternal, by Whom and in Whom are all things, visible and invisible.

7. From the Eternal they flow, to the Eternal they return. The spirit to Spirit, soul to Soul, mind to Mind, sense to Sense, life to Life, form to Form, dust to Dust.

8. In the beginning God willed and there came forth the beloved Son, the divine Love, and the beloved Daughter, the holy Wisdom, equally proceeding from the One Eternal Fount; and of these are the generations of the Spirits of God, the Sons and Daughters of the Eternal.

9. And These descend to earth, and dwell with men and teach them the ways of God, to love the laws of the Eternal, and obey them, that in them they may find salvation.

10. Many nations have seen their day. Under divers names have they been revealed to them, and they have rejoiced in their light; and even now they come again unto you, but Israel receiveth them not.

11. Verily I say unto you, my twelve whom I have chosen, that which hath been taught by them of old time is true—though corrupted by the foolish imaginations of men.

12. Again, Jesus spake unto Mary Magdalene saying, It is written in the law, Whoso leaveth father or mother, let him die the death. Now the law speaketh not of the parents in this life, but of the Indweller of light which is in us unto this day.

13. Whoso therefore forsaketh Christ the Saviour, the Holy law, and the body of the Elect, let them die the death. Yea, let them be lost in the outer darkness, for so they willed and none can hinder.

LECTION 64. -The occult teaching, in this discourse, of Jesus to his twelve has been handed down in spirit through the ages, but the world is blind and perceives not. See the same teaching in "New Light on Old Truths," founded on this Scripture.

LECTION 64. 8. -Beneath this profound saying of the Ghost Physician, the student cannot fail to notice the intimate and correct knowledge of the human frame, underlying the spiritual truth, which he enunciated. This knowledge has been claimed by science only some centuries later. The inner self- "alternate sex," in every man and woman, which occasionally manifests itself in the dream state, seems to be no mystery to him. (*See* G. Leland on " The Alternative Sex"; Welby, London).

Lection 65

The Last Anointing by Mary Magdalene.

1. NOW, on the evening of the Sabbath before the Passover, as Iesus was in Bethany he went to the house of Simon the leper, and there they made him a supper, and Martha served while Lazarus was one of them that sat at table with him.

2. And there came Mary called Magdalene, having an alabaster box of ointment of spikenard, very precious and costly, and she opened the box and poured the ointment on the head of Jesus, and anointed his feet, and wiped them with the hair of her head

3. Then said one among his disciples, Judas Iscariot, who was to betray him, Why is this waste of ointment which might have been sold for three hundred pence and given to the poor? And this he said not that he cared for the poor but because he was filled with jealousy and greed, and had the bag, and bare what was put therein. And they murmured against her.

4. And Jesus said, Let her alone, why trouble ye her? for she hath done all she could; yea, she hath wrought a good work on me. For ye have the poor always with you, but me ye have not always. She hath anointed my body for the day of my burial.

5. And verily, I say unto you, wheresoever this Gospel shall be preached in the whole world there shall also be told this that she hath done for a memorial of her.

6. Then entered Satan into the heart of Judas Iscariot and he went his way and communed with the chief priests and captains how he might betray him. And they were glad and covenanted with him for thirty pieces of silver, the price of a slave, and he promised them, and after that sought opportunity to betray him.

7. And at that time Iesus said to his disciples Preach ye unto all the world, saying, Strive to receive the mysteries of Light, and enter into the Kingdom of Light, for now is the accepted time and now is the day of Salvation.

8. Put ye not off from day to day, and from cycle to cycle and eon to eon, in the belief, that when ye return to this world ye will succeed in gaining the mysteries, and entering into the Kingdom of Light.

9. For ye know not when the number of perfected souls shall be filled up, and then will be shut the gates of the Kingdom of Light, and from hence none will be able to come in thereby, nor will any go forth.

10. Strive ye that ye may enter while the calls is made, until the number of perfected souls shall be sealed and complete, and the door is shut.

LECTION 65. 2. -It has been supposed by some, and not without some reason from the words of the Gospel, that envy and jealousy, and not greed of money, were the cause of Iudas' treachery, because he desired Mary Magdalene, and she had given all her love and devotion to her Master. This inner feeling seems to be concealed beneath the cloak of zeal for the poor. "From that hour he sought to betray him." It is as probable that all three motives urged him, as they do the multitudes nowadays, who grudge magnificence of architecture, music, etc., under the cloak of "unity." "These things ought ye to have done and not left the other undone." But such show their hypocrisy by their reckless contributions to war, and to all manner of pleasures and amusements and luxuries which minister to self. By the spirit of Iudas Iscariot are all such led and dominated.

Lection 66

Jesus Again Teacheth His Disciples

1. AGAIN Jesus taught them saying, God hath raised up witnesses to the truth in every nation and every age, that all might know the will of the Eternal and do it, and after that, enter into the kingdom, to be rulers and workers with the Eternal,

2. God is Power, Love and Wisdom, and these three are One. God is Truth, Goodness and Beauty, and these three are One.

3. God is Justice, Knowledge and Purity, and these three are One. God is Splendour, Compassion and Holiness, and these three are One.

4. And these four Trinities are One in the hidden Deity, the Perfect, the Infinite, the Onely.

5. Likewise in every man who is perfected, there are three persons, that of the son, that of the spouse. and that of the father, and these three are one.

6. So in every woman who is Perfected are there three persons, that of the daughter, that of the bride, and that of the mother and these three are one; and the man and the woman are one, even as God is One

7. Thus it is with God the Father-Mother, in Whom is neither male nor female and in Whom is both, and each is threefold, and all are One in the hidden Unity.

8. Marvel not at this, for as it is above so it is below, and as it is below so it is above, and that which is on earth is so, because it is so in Heaven.

9. Again I say unto you, I and My Bride are one, even as Maria Magdalena, whom I have chosen and sanctified unto Myself as a type, is one with Me; I and My Church are One. And the Church is the elect of humanity for the salvation of all.

10. The Church of the first born is the Maria of God. Thus saith the Eternal, She is My Mother and she hath ever conceived Me, and brought Me forth as Her Son in every age and clime. She is My Bride, ever one in Holy Union with Me her Spouse. She is My Daughter, for she hath ever issued and proceeded from Me her Father, rejoicing in Me.

11. And these two Trinities are One in the Eternal, and are strewn forth in each man and woman who are made perfect, ever being born of God, and rejoicing in light, ever being lifted up and made one with God, ever conceiving and bringing forth God for the salvation of the many.

12. This is the Mystery of the Trinity in Humanity, and moreover in every individual child of man must be accomplished the mystery of God, ever witnessing to the light, suffering for the truth, ascending into Heaven, and sending forth the Spirit of Truth And this is the path of salvation, for the kingdom of God is within.

13. And one said unto him, Master, when shall the kingdom come? And he answered and said, When that which is without shall be as that which is within, and that which is within shall be as that which is without, and, the male with the female, neither male nor female, but the two in One. They who have ears to hear, let them hear.

Lection 67

Entry Into Jerusalem

1. NOW on the first day of the week when they came nigh to Jerusalem, unto Bethage and Bethany, at the Mount of Olives, he sendeth forth two of his disciples, and saith unto them, Go your way into the village over against you, and as soon as you be entered into it, ye shalt find an ass tied, whereon never man sat, loose him and bring him.

2. And if any say unto you, Why do ye this? say ye that the Lord hath need of him, and straightway they will send him hither.

3. And they went their way and found the ass tied without in a place where two ways met, and they loosed him. And certain of them that stood there said unto them, What do ye, loosing the colt? And they said unto them, even as Iesus had commanded. And they let them go.

4. And they brought the ass to Iesus, and cast their garments upon him, and he sat upon the ass. And many spread their garments in the way, and others cut down branches off the trees and strewed them in the way.

5. And they that went before, and they that followed cried, saying, Hosanna, Blessed art thou who comest in the name of Jova: Blessed be the Kingdom of our ancestor David, and blessed be thou that comest in the name of the Highest: Hosanna in the highest.

6. AND Iesus entered into Jerusalem and into the Temple, and when he had looked round about upon all things, he spake this parable unto them, saying—

7. When the Son of man shall come in his glory and all the holy angels with him, then shall he sit upon the throne of his glory. And before him shall be gathered all nations, and he shall separate them one from another, as a shepherd divideth his sheep from the goats. And he shall set the sheep on his right hand, but the goats on the left.

8. Then shall the King say unto them on his right hand, Come ye blessed of my Parent, inherit the kingdom prepared for you from the foundation of the world. For I was an hungered and ye gave me food. was thirsty and ye gave me drink. I was a stranger anal ye took me in. Naked and ye clothed me. I was sick and ye visited me. I was in prison and ye came unto me.

9. Then shall the righteous answer him, saying, Lord, when saw we thee an hungered and fed thee? Or thirsty and gave thee drink? when saw we thee a stranger and took thee in? or naked and clothed thee? Or when saw we thee sick, or in prison and came unto thee ?

10. And the King shall answer and say unto them, Behold, I manifest myself unto you, in all created forms; and verily I say unto you, Inasmuch as ye have done it unto the least of these my brethren, ye have done it unto me.

11. Then shall he say also unto them on his left hand, Depart from me ye evil souls into the eternal fires which ye have prepared for yourselves, till ye are purified seven times and cleansed from your sins.

12. For I was an hungered and ye gave me no food, I was thirsty and ye gave me no drink. I was a stranger and ye took me not in, naked and ye clothed me not, sick and in prison and ye visited me not.

13. Then shall they also answer him, saying, Lord, when saw we thee an hungered, or athirst, or a stranger, or naked, or in prison, and did not minister unto thee ?

14. Then shall he answer them, saying, Behold I manifest myself unto you, in all created forms, and Verily I say unto you, Inasmuch as ye did it not to the least of these, my brethren, ye did it not unto me.

15. And the cruel and the loveless shall go away into chastisement for ages, and if they repent not, be utterly destroyed; but the righteous and the merciful, shall go into life and peace everlasting.

Lection 68

The Householder And The Husbandmen. Order Out Of Disorder.

1. AND Jesus said, Hear another parable: There was a certain householder, who planted a vineyard, and hedged it round about and digged a winepress in it, and built a tower, and let it out to husbandmen and went into a far country.

2. And when the time of the ripe fruits drew near, he sent his servants to the husbandmen that they might receive the fruits of it. And the husbandmen took his servants and beat one, and stoned another, and killed another.

3. Again he sent other servants, more honourable than the first, and they did unto them likewise. But last of all he sent unto them his son, saying, They will reverence my son.

4. But when the husbandmen saw the son, they said among themselves. This is the heir, come let us kill him, and let us seize on his inheritance. And they caught him and cast him out of the vineyard and slew him.

5. When the lord of the vineyard cometh what will he do unto those husbandmen? They say unto him, He will miserably destroy those wicked men and will let out his vineyard to other husbandmen, which shall render him the fruits in their seasons.

6. Jesus saith unto them, Did ye never read in the scriptures, The Stone which the builders rejected, the same is become the head of the Pyramid? this is the Lord's doing and it is marvellous in our eyes?

7. Therefore say I unto you, The kingdom of God shall be taken from you and given to a nation bringing forth the fruits thereof. And whosoever shall fall on this Stone shall be broken, but on whomsoever it shall fall, it will grind them to powder.

8. And when the chief priests and Pharisees had heard his parables, they perceived that he spake of them. But when they sought to lay hands on him they feared the multitude, because they took him for a prophet.

9. And the disciples asked him afterwards the meaning of this parable, and he said unto them, The vineyard is the world, the husbandmen are your priests, and the messengers are the servants of the good Law, and the Prophets.

10. When the fruits of their labour are demanded of the priests, none are given, but they evilly treat the

messengers who teach the truth of God, even as they have done from the beginning.

11. And when the Son of Man cometh, even the Christ of God, they gather together against the Holy One, and slay him, and cast him out of the vineyard, for they have not wrought the things of the Spirit, but sought their own pleasure and gain, rejecting the holy Law.

12. Had they accepted the Anointed One, who is the corner stone and the head, it would have been well with them, and the Building would have stood, even as the Temple of God inhabited by the Spirit.

13. But the day will come when the Law which they reject shall become the head stone, seen of all, and they who stumble on it shall be broken, but they who persist in disobedience shall he ground to pieces.

14. For to some of the angels God gave dominion over the course of this world, charging them to rule in wisdom. in justice and in love. But they have neglected the commands of the Most High, and rebelled against the good order of God. Thus cruelty and suffering and sorrow have entered the world, till the time the Master returns, and taketh possession of all things, and calleth his servants to account.

15. AND he spake another parable, saying: A certain man had two sons, and he came to the first and said, Son, go work today in my vineyard, and he answered and said, I will not, but afterwards he repented and went. And he came to the second and said likewise, and he answered and said, I go, sir, and went not. Whether of them twain did the will of his father?

16. They say unto him, The first, and Iesus saith unto them, Verily I say unto you, That the publicans and harlots go into the kingdom of God before you. For John came unto you in the way of righteousness and ye believed him not, but the taxgatherers and the harlots believed him, and ye, when ye had seen it, repented not afterwards, that ye might believe him.

17. AND the Lord gathered together all his disciples in a certain place. And he said unto them, Can ye make perfection to appear out of that which is imperfect? Can ye bring order out of disorder? And they said, Lord, we cannot.

18. And he placed them according to the number of each in a four-square order, each side lacking one of twelve (and this he did, knowing who should betray him, who should be counted one of them by man, but was not of them)

19. The first in the seventh rank from above in the middle, and the last in the seventh from below, and him that was neither first nor last did he make the Centre of all, and the rest according to a Divine order did he place them, each finding his own place, so those which were above, were even as those which were below, and the left side was equal to the right side, and the right side to the left, according to the sum of their numbers.

20. An he said, See you how ye stand? I say unto you, In like manner is the order of the kingdom, and the One who ruleth all is in your midst, and he is the centre, and with him are the hundred and twenty, the elect of Israel, and after them cometh the hundred and forty and four thousand, the elect of the Gentiles, who are their brethren.

LECTION 68. -Again, studying these "dark sayings" so difficult to understand, recourse has been had to certain figures known to the early Gnostics. *(See* the "Squares and Circles" by the Editor of this Gospel.) The Magic Square of 11 has been found wonderfully to explicate, symbolically at least, to the mystical, the meaning of the passage. The form exactly illustrates what the Lord in symbol taught to his disciples of the bringing forth of order out of disorder, perfection out of imperfection, and out of deficiency fulness. Compare the Magic Square given below with the natural square which anyone can form by writing the numbers in consecutive order; the result is at once seen, and may help to arrive at the meaning of this very mystical passage:- **See Square**

Lection 69

The Christ Within, The Resurrection And The Life.

1. As Iesus sat by the west of the Temple with his disciples, behold there passed some carrying one that was dead to burial, and a certain one said unto him, Master, if a man die, shall he live again?

2. And he answered and said, I am the resurrection and the life, I am the Good, the Beautiful, the True, if

a man believe in me he shall not die, but live eternally. As in Adam all die, so in the Christ shall all be made alive. Blessed are the dead who die in me, and are made perfect in my image and likeness, for they rest from their labours and their works do follow them. They have overcome evil, and are made Pillars in the Temple of my God, and they go out no more, for they rest in the Eternal.

3. For them that have done evil there is no rest, but they go out and in, and suffer correction for ages, till they are made perfect. But for them that have done good and attained unto perfection, there is endless rest and they go into life everlasting. They rest in the Eternal.

4. Over them the repeated death and birth have no power, for them the wheel of the Eternal revolves no more, for they have attained unto the Centre, where is eternal rest, and the centre of all things is God.

5. AND one of the disciples asked him, How shall a man enter into the Kingdom? And he answered and said, If ye make not the below as the above, and the left as the right, and the behind as the before, entering into the Centre and passing into the Spirit, ye shall not enter into the Kingdom of God.

6. And he also said, Believe ye not that any man is wholly without error for even among the prophets. and those who have been initiated into the Christhood, the word of error has been found. But there are a multitude of error which are covered by love.

7. AND now when the eventide was come, he went out unto Bethany with the twelve. For there abode Lazarus and Mary and Martha whom he loved.

8. And Salome came unto him, and asked him, saying, Lord, how long shall death hold sway? And he answered, So long as ye men inflict burdens and ye woman bring forth, and for this purpose I am come, to end the works of the heedless..

9. And Salome saith unto him, Then I have done well in not bringing forth. And the Lord answered and said Eat of every pasture which is good, but of that which hath the bitterness of death, eat not.

10. And when Salome asked when those things of which she enquired should be known, the Lord said, When ye shall tread upon the vesture of shame and rise above desire; when the two shall be one, and the male with the female shall be neither male nor female.

11. And again, to another disciple who asked, When shall all obey the law? Iesus said, When the Spirit of God shall fill the whole earth and every heart of man and of woman.

12. I cast the law into the earth and it took root and bore in due time twelve fruits for the nourishment of all. I cast the law into the water and it was cleansed from all defilements of evil. I cast the law into the fire, and the gold was purged from all dross. I cast the law into the air, and it was made alive by the Spirit of the Living One that filleth all things and dwelleth in every heart.

13. And many other like sakings he spake unto them who had ears to hear, and an understanding mind. But to the multitude they were dark sayings.

LECTION 69. 8, 9. -This saying of Jesus is very difficult to the popular mind, as apparently reversing the original injunction in Gen.1-3, "Be ye fruitful and multiply." To understand this, it must be borne in mind that the promise of a Messiah to redeem the world, has, from the earliest times, begotten in the woman of the Hebrew nation, to whom it was specially given, that insatiable desire for offspring, each woman thinking of herself as the possible mother of Him who was to come and save. Iesus, the true Prophet, seeing the tendency to the propagation of the unfit (as we now see) to bring want, misery, squalor, vice and crime, through the inability of most parents to bring them up as they should be, owing to the curse of competition and greed, here proclaimed to Salome, in answer to her query, that he would reverse all this tendency among his followers, and through them, extend this reversal to mankind at large, for He, the desire and hope of nations, having come, there was no longer any supposed necessity or reason for such increase, of which He, the Prophet of God, fully foresaw the evil, in the ages to come, as we now fully experience it.

Lection 70

Jesus Rebukes Peter For His Haste.

1. NOW on the morrow as they were coming from Bethany, Peter was hungry, and perceiving a fig tree afar off having leaves thereon, he came if happily he might find fruit thereon, and when he came he found nothing but leaves, for the time of figs was not yet.

2. And Peter was angry and said unto it, Accursed tree, no man eat fruit of thee hereafter for ever. And some of the disciples heard of it.

3. And the next day as Jesus and his disciples passed by, Peter said unto Jesus, Master, behold, the fig tree which I cursed is green and flourishing, wherefore did not my word prevail?

4. Iesus said unto Peter, Thou knowest not what spirit thou art of. Wherefore didst thou curse that which God hath not cursed? And Peter said, Behold Lord I was a hungered, and finding leaves and no fruit, I was angry, and I cursed the tree.

5. And Jesus said, Son of Jonas knewest thou not that the time of figs was not yet? Behold the corn which is in the field which groweth according to its nature first the green shoot, then the stalk, then the ear—would thou be angry if thou camest at the time of the tender shoot or the stalk, and didst not find the corn in the ear? And wouldst thou curse the tree which, full of buds and blossoms, had not yet ripe fruit?

6. Verily Peter I say unto thee, one of my twelve will deny me thrice in his fear and anger with curses, and swear that he knows me not, and the rest will forsake me for a season.

7. But ye shall repent and grieve bitterly, because in your heart ye love me, and ye shall be as an Altar of twelve hewn stones, and a witness to my Name, and ye shall be as the Servants of servants, and the keys of the Church will I give unto you, and ye shall feed my sheep and my lambs and ye shall be my vice-gerents upon earth.

8. But there shall arise men amongst them that succeed you, of whom some shall indeed love me even as thou, who being hotheaded and unwise, and void of patience, shall curse those whom God hath not cursed, and persecute them in their ignorance, because they cannot yet find in them the fruits they seek.

9. And others being lovers of themselves shall make alliance with the kings and rulers of the world, and seek earthly power, and riches, and domination, and put to death by fire and sword those who seek the truth, and therefore are truly my disciples.

10. And in their days I Iesus shall be crucified afresh and put to open shame, for they will profess to do these things in my Name. And Peter said, Be it far from thee Lord.

11. And Iesus answered, As I shall be nailed to the cross, so also shall my Church in those days, for she is my Bride and one with me. But the day shall come when this darkness shall pass away, and true Light shall shine.

12. And one shall sit on my throne, who shall be a Man of Truth and Goodness and Power, and he shall be filled with love and wisdom beyond all others, and shall rule my Church by a fourfold twelve and by two and seventy as of old, and that only which is true shall he teach.

13. And my Church shall be filled with Light, and give Light unto all nations of the earth, and there shall be one Pontiff sitting on his throne as a King and a Priest.

14. And my Spirit shall be upon him and his throne shall endure and not be shaken, for it shall be founded on love and truth and equity, and light shall come to it, and go forth from it, to all the nations of the earth, and the Truth shall make them free.

LECTION 70. 1-5. -Long has Jesus suffered reproach, for those words so falsely attributed to him, in place of the impulsive Peter, who spoke them, and with whose character they were in full harmony.

The Gospel of the Holy Twelve

Translated from the original Aramaic
by Rev. G.J.R. Ouseley

Section 8, Lections 71 thru 80

Lection 71

The Cleansing Of The Temple

1. AND the Jews' Passover was at hand, and Jesus went up again from Bethany into Jerusalem. And he found in the temple those that sold oxen and sheep and doves, and the changers of money sitting.

2. And when he had made a scourge of seven cords, he drove them all out of the temple and loosed the sheep and the oxen, and the doves, and poured out the changers' money, and overthrew the tables;

3. And said unto them, Take these things hence; make not my Father's House an House of merchandise. Is it not written, My House is a House of prayer, for all nations? but ye have made it a den of thieves, and filled it with all manner of abominations.

4. And he would not suffer that any man should carry any vessel of blood through the temple, or that any animals should be slain. And the disciples remembered that it was written, Zeal for thine house hath eaten me up.

5. Then answered the Jews, and said unto him, What sign shewest thou unto us, seeing that thou doest these things? Iesus answered and said unto them, Again I say unto you, Destroy this temple, and in three days I will raise it up.

6. Then said the Jews, Forty and six years was this temple in building and wilt thou rear it up in three days? But he spake of the temple of his Body.

7. When therefore he was risen from the dead, his disciples remembered that he had said this unto them; and believed the scripture and the word which Iesus had said.

8. But the scribes and the priests saw and heard, and were astonished and sought how they might destroy him, for they feared him, seeing that all the people were attentive to his doctrines.

9. And when even was come he went out of the city. For by day he taught in the Temple and at night he went out and abode on the Mount of Olives, and the people came early in the morning to hear him in the Temple courts.

10. Now when he was in Jerusalem at the passover, many believed in his Name, when they saw the miracles which he did.

11. But Iesus did not commit himself unto them, because he knew all men. And needed not that any should testify of man; for he knew what was in man.

12. And Iesus seeing the passover night was at hand, sent two of his disciples, that they should prepare the upper room where he desired to eat with his twelve, and buy such things as were needful for the feast which he purposed thereafter.

LECTION 71. 3. - Often translated as slaughterhouse in text having the characteristics of language past and not surviving chiefly on behalf of the revisionists alone.

LECTION 71. 1-4. -Twice the Lord is said to have performed this symbolic act. Surely, at his return, it will be his first work! For since the first ages till now the spirit of the world ruleth, and mammon is dominant, and every kind of wickedness in the name of religion, zeal for purity, etc.

Lection 72

The Many Mansions In The One House

1. AND as Iesus sat with his disciples in the Garden of Gethsemane he said unto them: Let not your heart be troubled; ye believe in God, believe also in me. In my parent's house are many mansions: if it were not so, I would have told you. I go to prepare a place for you. And if I go and prepare a place for you, I will come again, and receive you unto myself; that where I am, there ye may be also. And whither I go ye know, and the way ye know.

2. Thomas said unto him, Lord, we know not whither thou goest; and how can we know the way? Iesus saith unto him, I am the Way, the Truth, and the Life: no man cometh unto the All Parent but by me, If ye had known me, ye should have known my Parent also: and from henceforth ye know and have seen my Parent.

3. Philip saith unto him, Lord, shew US the All-Parent and it sufficeth us. Iesus saith unto him, Have I been so long time with

you, and yet hast thou not known me, Philip? he that hath seen me hath seen the All-Parent; and how sayest thou then, Shew us the All-Parent? Believest thou not that I am in the All-Parent, and the All-Parent in me? the words that I speak unto you I speak not of myself: but the All-Parent who dwelleth in me doeth the works.

4. Believe me, that I am in the All-Parent and the All-Parent in me: or else, believe me for the very works' sake. Verily, verily, I say unto you, They who believe on me, the works that I do shall they do also; and greater works than these shall they do; because I go unto my Parent.

5. And whatsoever ye shall ask in my Name, that will I do, that the All-Parent may be glorified in the Son and Daughter of Man. If ye shall ask anything in my Name, I will do it.

6. If ye love me, keep my commandments. And I will pray the All-Parent, Who shall give you another Comforter, to abide with you for ever; even the Spirit of truth. whom the world cannot receive, because it seeth not, neither knoweth, but ye know; for the Spirit dwelleth with you, and shall be in you.

8. They who have my commandments, and keep them, these are they who love me; and they that love me shall be loved of my Parent, and I will love them and will manifest myself to them.

9. Judas saith unto him, Lord, how is it that thou wilt manifest thyself unto us, and not unto the world? Iesus answered and said unto him, If any love me, they will keep my words: and the Holy One will love them and we will come unto them, and make our abode with them.

10. They that love me not keep not my sayings: and the word which ye hear is not mine, but the All-Parent's who sent me. These things have I spoken unto you, being yet present with you. But the Comforter, who is my Mother, Holy Wisdom, whom the Father will send in my name, she shall teach you all things, and bring all things to your remembrance, whatsoever I have said unto you.

11. Peace I leave with you, my peace I give unto you: not as the world giveth, give I unto you. Let not your heart be troubled, neither let it be afraid. Ye have heard how I said unto you, I go away, and come again unto you. If ye loved me ye would rejoice, because I said, I go unto the All-Parent: for the All-Parent is greater than I.

12. And now I have told you before it come to pass, that, when it is come to pass, ye may believe. Hereafter I will not talk much with you; for the prince of this world cometh, and hath nothing in me.

13. But that the world may know that I love the All-Parent; as the All-Parent gave me commandment, even so I do. Even unto the end.

LECTION 72. 1. -In the language of the Churches of this day, there is but one mansion in the Father's house, and *that* is claimed by each of over 300 different sects as its own, and all outside are damned, not for their evil deeds, but because they cannot see as their rulers profess to see.

Lection 73

The True Vine

1. AFTER these things Iesus spake saying unto them: I am the true vine, and my Parent is the vinedresser. Every branch in me that beareth not fruit is taken away: and every branch that beareth fruit, is purged that it may bring forth more fruit.

2. Abide in me, and I in you. As the branch cannot bear fruit of itself, except it abide in the vine; no more can ye, except ye abide in me. I am the tree, ye are the branches: Whoso abide in me and I in them, the same bring forth much fruit; for without me ye do nothing.

3. If any abide not in me, they are cast forth as useless branches, and they wither away; and men gather them, and cast them into the fire, and they are burned. If ye abide in me, and my words abide in you, ye shall ask what ye will, and it will be done unto you.

4. Verily, I am the true Bread which cometh down out of Heaven, even the Substance of God which is one with the Life of God. And, as many grains are in one bread, so are ye, who believe, and do the will of my Parent, one in me. Not as your ancestors did eat manna and are dead; but they who eat this Bread shall live for ever.

5. As the wheat is separated from the chaff, so must ye be separated from the falsities of the world; yet must ye not go out of the world, but live separate in the world, for the life of the world.

6. Verily, verily, the wheat is parched by fire, so must ye my disciples pass through tribulations. But rejoice ye: for having suffered with me as one body ye shall reign with me in one body, and give life to the world.

7. Herein is my Parent glorified, that ye bear much fruit; so shall ye be my disciples. As the All-Parent hath loved me, so have I loved you: continue ye in my love. If ye keep my commandments, ye shall abide in my love; even as I have kept my Parent's commandments, and abide in the spirit of love.

8. These things have I spoken unto you, that my joy might remain in you, and that your joy might be full. This is my commandment, That ye love one another, as I have loved you. Greater love hath no man than this, that a man lay down his life for his friend Ye are my friends, if ye do whatsoever I command you.

9. Henceforth I call you not servants; for the servant knoweth not what his lord doeth: but I have called you friends; for all things that I have heard of my Parent I have made known unto you. Ye have not chosen me, but I have chosen you, and ordained you, that ye should remain: that whatsoever ye shall ask of the All-Parent in my Name, ye may receive.

10. These things I command you, that ye love one another and all the creatures of God. If the world hate you, ye know that it hated me before it hated you. If ye were of the world, the world would love its own: but because ye are not of the world, but I

have chosen you out of the world' therefore the world hateth you.

11. Remember the word that I said unto you, The servant is not greater than his lord. If they have persecuted me, they will also persecute you; if they have kept my saying, they will keep yours also. But all these things will they do unto you for my Name's sake, because they know not him that sent me.

12. If I had not come and spoken unto them, they had not had sin: but now they have no cloke for their sin. He that hateth me hateth my Parent also. If I had not done among them the works which none other man did, they had not had sin: but now have they, have seen and hated both me and my Parent. But this cometh to pass, that the word might be fulfilled that is written in their law, They hated me without a cause.

13. But when the Comforter is come, Whom I will send unto you from the All Parent, even the Spirit of truth, which proceedeth from the Father and the Mother the same shall testify of me: And ye also shall bear witness, because ye have been with me from the beginning.

LECTION 73. 1-6. -"I am the true Vine, ye are the branches" -in unity with the stem by the continual possession of the One Life, not by *mere* external unity, valuable as this is, and certainly not by a dead uniformity of opinion in all things. *"Tot homines tut sententice."*

Lection 74

Iesus Foretelleth Persecutions

1. THESE things have I spoken unto you that ye should be forewarned, They shall put you out of the synagogues; yea, the time cometh, that whosoever killeth you will think that they do God's service. And these things will they do unto you, because they have not known the All Parent, nor me.

2. But these things have I told you, that when the time shall come, ye may remember that I told you of them. And these things I said not unto you at the beginning, because I was with you. But now I go my way to my Parent that sent me; and none of you asketh me, Whither goest thou? But because I have said these thing unto you, sorrow hath filled your heart.

3. Nevertheless I tell you the truth; It is expedient for you that I go away; for if I go not away, the Comforter will not come unto you; but if I depart, I will send my Spirit unto you. And when the Spirit is come, the world shall be reproved of sin and of righteousness, and of judgement.

4. Of sin, because they believe not on me; of righteousness, because I go to my Father, and ye see me no more; of judgement, because the prince of this world is judged.

5. I have yet many things to say unto you, but ye cannot bear them now. Howbeit when the Spirit of Truth is come, she will guide you into all truth: and the same will shew you things to come and shall glorify me: for the same shall receive of mine, and shall shew it unto you.

6. All things that my Parent hath are mine: therefore said I, that the Comforter shall take of mine and shall shew it unto you. A little while, and ye shall not see me: and again, a little while, and ye shall see me, because I go to the All-Parent. Then said some of his disciples among themselves, What is that he saith unto us, A little while, and ye shall not see me: and again, a little while, and ye shall see me; and, Because I go to the All-Parent?

7. Now Iesus knew that they were desirous to ask him, and said unto them, Do ye enquire among yourselves of that I Said, A little while, and ye shall see me? Verily, verily, I say unto you, That ye shall weep and lament, but the world shall rejoice: and ye shall be sorrowful, but your sorrow shall be turned into joy.

8. A woman when she is in travail hath sorrow, because her hour is come: but as soon as she is delivered of the child, she remembereth no more the anguish, for joy that a man is born into the world. And ye now therefore have sorrow; but I will see you again, and your heart shall rejoice, and your joy no man taketh from you.

9. And in that day ye shall ask me nothing. Verily, verily, I say unto you, Whatsoever ye shall ask my Parent in my name, ye will receive. Hitherto have ye asked nothing in my name: ask and ye shall receiveth that your joy may be full. These things have I spoken unto you in proverbs; but the time cometh, when I shall no more speak unto you in a mystery, but I shall shew you plainly of the All-Parent.

10. At that day ye shall ask in my name: and I say not unto you, that I will pray my Parent for you; For the All-Parent in truth loveth you, because ye have loved me, and have believed that I came out from God. I came forth from God, and am come into the world; again, I leave the world, and go unto my God.

11. His disciples said unto him, Lo, now speakest thou plainly, and speakest no mystery. Now are we sure that thou knowest all things, and needest not that any man should ask thee: by this we believe that thou comest forth from God.

12. Iesus answered them, Do ye now believe? Be hold, the hour cometh, yea, is now come, that ye shall be scattered, every man to his own home, and shall leave me alone: and yet I am not alone, because the Father is with me.

13. These things I have spoken unto you, that in me ye might have peace. In the world ye shall have tribulation: but be of good cheer; I have overcome the world. Arise, let us go hence.

Lection 75

The Last Paschal Supper

1. AND at evening the Master cometh into the house, and there are gathered with him the Twelve and their fellows; Peter and Jacob and Thomas and John and Simon and Matthew and Andrew and Nathanael and James and Thaddeus and Jude and Philip and their companions (and there was also Judas Iscariote, who by men was numbered with the twelve, till the time when he should be manifested).

2. And they were all clad in garments of white linen, pure and clear, for linen is the righteousness of the saints; and each had the colour of his tribe. But the Master was clad in his pure white robe, over all, without seam or spot.

3. And there arose contention among them as to which of them should be esteemed the greatest, wherefore he said unto them, He that is greatest among you let him be as he that doth serve.

4. And Iesus said, With desire have I desired to eat this Passover with you before I suffer. and to institute the Memorial of my Oblation for the service and salvation of all. For behold the hour cometh when the Son of man shall be betrayed into the hands of sinners.

5. And one of the twelve said unto him, Lord, is it I ? And he answered, He to whom I give the sop the same is he.

6. And Iscariot said unto him, Master, behold the unleaven bread, the mingled wine and the oil and the herbs, but where is the lamb that Moses commanded? (for Judas had bought the lamb, but Iesus had forbidden that it should be killed).

7. And John spake in the Spirit, saying, Behold the Lamb of God, the good Shepherd which giveth his life for the sheep. And Judas was troubled at these words, for he knew that he should betray him. But again Judas said, Master, is it not written in the law that a lamb must be slain for the passover within the gates?

8. And Iesus answered, If I am lifted up on the cross then indeed shall the lamb be slain; but woe unto him by whom it is delivered into the hands of the slayers; it were better of him had he not been born.

9. **Verily I say unto you, for this end have I come into the world, that I may put away all blood offerings and the eating of the flesh of the beasts and the birds that are slain by men.**

10. In the beginning, God gave to all, the fruits of the trees, and the seeds, and the herbs, for food; but those who loved themselves more than God, or their fellows, corrupted their ways, and brought diseases into their bodies, and filled the earth with lust and violence.

11. Not by shedding innocent blood, therefore, but by living a righteous life, shall ye find the peace of God. Ye call me the Christ of God and ye say well, for I am the Way, the Truth and the Life.

12. Walk ye in the Way, and ye shall find God. Seek ye the Truth, and the Truth shall make you free. Live in the Life, and ye shall see no death. All things are alive in God, and the Spirit of God filleth all things.

13. Keep ye the commandments. Love thy God with all thy heart, and love thy neighbour as thyself. On these hang all the law and the prophets. And the sum of the law is this—Do not ye unto others as ye would not that others should do unto you. Do ye unto others, as ye would that others should do unto you.

14. Blessed are they who keep this law, for God is manifested in all creatures. All creatures live in God, and God is hid in them.

15. After these things, Iesus dipped the sop and gave it to Judas Iscariot, saying, What thou doest, do quickly. He then, having received the sop, went out immediately, and it was light.

16. And when Judas Iscariot had gone out, Iesus said, Now is the Son of man glorified among his twelve, and God is glorified in him. And verily I say unto you, they who receive you receive me, and they who receive me receive the Father-Mother Who sent me, and ye who have been faithful unto the truth shall sit upon twelve thrones, judging the twelve tribes of Israel.

17. And one said unto him, Lord, wilt thou at this time restore the kingdom unto Israel? And Iesus said, My kingdom is not of this world, neither are all Israel which are called Israel.

18. They in every nation who defile not themselves with cruelty, who do righteousness, love mercy, and reverence all the works of God, who give succour to all that are weak and oppressed—the same are the Israel of God.

LECTION 75. 1. -Jacob is the same as James -called "the great." Nathanael is Bartholomew. There is no proof that Jude was the same with Thaddeus, as is alleged by some. The number at first seems to have been twelve exclusive, or thirteen (to the world's eye) including Iudas Iscariot, till he should manifest his falsity by his treachery, when he went out directly before the holy supper, leaving Iesus with the twelve -the complete number of Apostleship, which, being even, admitted of no one among them being "Master," save Iesus, who was over them.

v. 2. -Whether the appearance of the Master and his disciples in symbolic festal garb may not have been seen only by the spiritual eye of some of the disciples or not, the lesson is the same. Reverence and love of beauty and order are to be seen in God's House -symbols of the glorious garments of that Being Who is the Eternal Mystery and Beauty manifest in all things.

vv. 15, 16. -That Twelve is the complete number of the Apostleship and that Iesus sat down *"with his twelve"* at the holy supper before his crucifixion, seems evident from the received gospels, and still more so, from the fragments lately brought to light. Iudas Iscariot appears then to have been *among* the twelve but not of them, therefore before the Eucharistic rite is celebrated "he goes out." If there were any ill omen at all about the number thirteen it would therefore be thirteen as the number of Apostles present, exclusive of the Master and Head. But to thirteen, inclusive of the presiding host, no ill omen could attach, but the reverse.

Lection 76

The Washing Of Feet.
New Commandment.
The Eucharistic Oblation.

1. AND the Paschal Supper being ended, the lights were kindled, for it was even. And Iesus arose from the table and laid aside his garment, and girded himself with a towel, and pouring water into a basin, washed the feet of each of the fourfold Twelve, and wiped them with the towel with which he was girded.

2. And one of them said, Lord, thou shalt not wash my feet. And Iesus said, If I wash thee not thou hast no part with me. And he answered, Lord, wash not my feet only, but my head and my hands.

3. And he said unto him, They who have come out of the bath, need not but to wash their feet, and they are clean every whit.

4. AND then putting on the overgarment of pure white linen without spot or seam, he sat at the table and said unto them, Know ye what I have done unto you? Ye call me Lord and Master, and if then your Lord and Master have washed your feet, ye ought also to wash one another's feet. For I have given this example, that as I have done unto you, so also should ye do unto others.

5. A new commandment I give unto you, that ye love one another and all the creatures of God. Love is the fulfilling of the law. Love is of God, and God is love. Whoso loveth not, knoweth not God.

6. Now ye are clean through the word which I have spoken unto you. By this shall all men know that ye are my disciples if ye have love one to another and shew mercy and love to all creatures of God, especially to those that are weak and oppressed and suffer wrong. For the whole earth is filled with dark places of cruelty, and with pain and sorrow, by the selfishness and ignorance of man.

7. I say unto you, Love your enemies, bless them that curse you, and give them light for their darkness and let the spirit of love dwell within your hearts, and abound unto all. And again I say unto you, Love one another, and all the creation of God And when he had finished, they said, Blessed be God.

8. Then he lifted up his voice, and they joined him, saying, As the hart panteth after the water brooks, so panteth my soul after thee, O God. And when they had ended, one brought unto him a censer full of live coals, and he cast frankincense thereon even the frankincense which his mother had given him in the day of his manifestation, and the sweetness of the odour filled the room.

9. Then Jesus, placing before him the platter, and behind it the chalice, and lifting up his eyes to heaven, gave thanks for the goodness of God in all things and unto all, and after that he took in his hands the unleavened bread, and blessed it; the wine likewise mingled with water and blessed it; chanting the Invocation of the Holy Name the Sevenfold, calling upon the thrice Holy Father-Mother in Heaven to send down the Holy Spirit and make the bread to be his body, even the Body of the Christ, and the fruit of the vine to be his Blood, even the Blood of the Christ, for the remission of sins and everlasting life, to all who obey the gospel.

10. Then lifting up the Oblation towards heaven, he said, The Son who is also the Daughter of man is lifted up from the earth, and I shall draw all men unto me; then it shall be known of the people that I am sent from God.

11. These things being done, Iesus spake these words, lifting his eyes to heaven. Abba Amma, the hour is come, Glorify thy Son that Thy Son may be glorified in thee.

12. Yea, Thou hast glorified me, Thou hast filled my heart with fire, Thou hast set lamps on my right hand and on my left, so that no part of my being should be without light. Thy Love shineth on my right hand and on my left, so that no part of my being should be without light. Thy Love shineth on my right hand, and Thy Wisdom on my left. Thy Love, Thy Wisdom, Thy Power are manifest in me.

13. I have glorified Thee on earth, I have finished the work Thou gavest me to do. Holy One, keep through Thy Name the Twelve and their fellows whom Thou hast given me, that they may be One even as we are One. Whilst I was with them in the world I kept them in Thy Name, and none of them is lost, for he who went out from us, was not of us, nevertheless, I pray for him that he may be restored. Father-Mother, forgive him, for he knoweth not what he doeth.

14. And now come I to Thee, and these things I speak in the world that they may have my joy fulfilled in themselves. I give them Thy word, and the world hath them, because they are not of the world, even as I am not of the world.

15. I pray not that Thou shouldst take them out of the world, but that Thou shouldst keep them from evil, whilst yet in the world, Sanctify them through Thy truth. Thy word is Truth. As thou sendest me into the world, so also I send them into the world, and for their sakes I sanctify myself, that they also may be sanctified through the Truth.

16. Neither pray I for these alone, but for all that shall be added to their number, and for the Two and Seventy also whom I sent forth, yea, and for all that shall believe in the Truth through Thy word, that they also may be one as Thou Most Holy art in me and I in Thee, that they may also be one in Thee, that the world may know that Thou hast sent me.

17. Holy Parent, I will also, that they whom Thou hast given me, yea all who live, be with me where I am, that they may partake of my glory which thou givest me, for Thou lovest me in all, and all in me, from before the foundations of the world.

18. The world hath not known Thee in Thy righteousness, but I know Thee, and these know that Thou hast sent me.

19. And I have declared unto them Thy Name that the love wherewith Thou hast loved me may be in them, and that from them it may abound, even unto all Thy creatures, yea, even unto all These words being ended, they all lifted up their voices with him, and prayed as he taught them, saying:

20. Our Father-Mother: Who art above and within. Hallowed be Thy sacred Name, in Biune Trinity. In Wisdom, Love and Equity Thy Kingdom come to all. Thy holy Will be done always, as in Heaven, so on Earth. Give us day by day to partake of Thy holy Bread, and the fruit of Thy living Vine. As we seek to perfect others, so perfect us in Thy Christ. Shew upon us Thy goodness, that to others we many shew the same. In the hour of trial, deliver us from evil.

21. For Thine are the Kingdom, the Power and the Glory: From the Ages of ages, Now, and to the Ages of ages. Amun.

22. THEN our Master taketh the holy Bread and breaketh it, and the Fruit of the Vine also, and mingleth it, and having blessed and hallowed both, and casting a fragment of the Bread into the Cup, he blessed the holy Union.

23. Then he giveth the bread which he had hallowed to his disciples saying, Eat ye, for this is my Body, even the Body of the Christ, which is given for the Salvation of the body and the soul.

24. Likewise he giveth unto them the fruit of the Vine which he had blessed saying unto them, Drink ye, for this is my Blood, even the Blood of the Christ which is shed for you and for many, for the Salvation of the Soul and the Body.

25. And when all had partaken, he said unto them, As oft as ye assemble together in my Name, make this Oblation for a Memorial of me, even the Bread of everlasting life and the Wine of eternal salvation' and eat and drink thereof with pure heart, and ye shall receive of the Substance and the Life of God, which dwelleth in me.

26. And when they had sung a hymn, Iesus stood up in the midst of his apostles, and going to him who was their Centre, as in a solemn dance, they rejoiced in him. And then he went out to the Mount of Olives, and his disciples followed him.

27. Now Judas Iscariot had gone to the house of Caiaphas and said unto him, Behold he has celebrated the Passover, within the gates, with the Mazza in place of the lamb. I indeed bought a lamb, but he forbade that it should be killed, and lo, the man of whom I bought it is witness.

28. And Caiaphas rent his clothes and said, Truly this is a Passover of the law of Moses. He hath done the deed which is worthy of death, for it is a weighty transgression of the law. What need of further witness? Yea, even now two robbers have broken into the Temple and stolen the book of the law, and this is the end of his teaching. Let us tell these things to the people who follow him, for they will fear the authority of the law.

29. And one that was standing by as Judas came out, said unto him, Thinkest thou that they will put him to death?

30. And Judas said, Nay, for he will do some mighty work to deliver himself out of their hands, even as when they of the synagogue in Capernaum rose up against him, and brought him to the brow of the hill that they might throw him down headlong, and did he not pass safely through their midst? He will surely escape them now also, and proclaim himself openly and set up the Kingdom whereof he spake.

LECTION 76. 4. -There are two other alternative versions of these circumstances of the last supper in the A. V .-First, that of St. John who, in the received version, expressly affirms that Iesus was crucified on the very day of the Passover and consequently the Eucharist was instituted the day before and not on the feast day Itself' and the Passover was on the morrow after the trial on the day of the crucifixion. Secondly, that of the three other gospels, which all affirm that the Eucharist was Instituted on the Passover the pascal lamb was slain. If the latter, it must be remembered that the Essenes (of whom Iesus was apparently one), were by Jewish regulation allowed a separate table at which no lamb or other flesh-meat was eaten, as they were vowed abstainers from blood sacrifices and the eating of flesh. If the former it was not the Passover at all, and Iesus was not bound as a Jew to eat of a lamb. In neither of these cases, therefore, was Jesus under the alleged necessity of killing a lamb and eating of flesh-meat in order to fulfill the law. In any case the causing of an innocent lamb to be killed and the eating of such is contrary to all that is known of the character of Iesus the Christ, whose tender love extends to all creatures. If Iesus was not an Essene, then nothing can be said against the accuracy of this version of the holy supper, and the charges brought against him in the account of the trial as now given by the Spirit.

v. 9. -"Bread," *i.e.* unleavened cakes of pure meal such as in use at the Passover. "Wine," here and through the Gospels, as used by Jesus and His disciples, means "the fruit of the Vine." which is pure wine mingled with four or two parts of pure water, the latter mystically representing the humanity, and the former the Divine Spirit. The strong fermented wine of modem use was never used on such festive occasions, nor even generally, except thus mingled with water. It is to be noted that the Saviour consecrated the Eucharist by Invocation of the Holy Spirit, and this has been faithfully followed by all Churches of the East, the words of institution being merely recited before, as a historical preamble, giving the authority for the action, and in no case as the words of consecration, according to the corrupt use of the West.

v. 13. -In the received Gospel Iudas is consigned to eternal perdition, but it appears rather that he who was all compassion and prayed for his murderers, prayed also for the man who was overmastered by his passions, blinded by envy, jealousy, greed of money, or, as some say, by desire to push matters to their conclusion, and procure some decisive miracle that would establish the claim of his Master to set up a temporal kingdom.

v. 26. -It is not stated whether there was any musical accompaniment, as is usual in the religious dances and processions of the East, but if so it was probably of the simplest, such as the Pipe, used on such occasions.

v. 27. -The Mazza, or unleavened cake, to which may the word "Mass" be traced as applied to the Eucharist, or "Breaking of Bread" -but preferable perhaps is the interpretation of "ite missa est " -the oblation (= prayer) is gone, *"sent up."*

v. 30. -Here was perhaps more probably the sole motive actuating Judas -his ambition- the desire to see a miracle, and the early sovereignty set up before the time.

Lection 77

The Agony In The Garden

1. AND as they went to the Mount of Olives, Iesus said unto them, All ye shall be offended because of me this night; for it is written, I will smite the shepherd, and the sheep of the flock shall be scattered abroad. But after I am risen again, I will go before you into Galilee.

2. Simon answered and said unto him, Though all men shall be offended because of thee, yet will I never be offended. And the Lord said, Simon, Simon, behold Satan hath desired to have you, that he may sift you as wheat. But I have prayed for thee that thy faith fail not; and when thou art converted, strengthen thy brethren.

3. And he said unto him, Lord, I am ready to go with thee, both unto prison and unto death. And Iesus said, I tell thee, Simon, the cock shall not crow this night, before that thou shalt thrice deny that thou knowest me.

4. Then cometh Iesus with them, having crossed the brook Kedron, unto the garden called Gethsemane, and saith unto the disciples, Sit ye here while I go and pray yonder. (Judas also, which betrayed him, knew the place, for Iesus ofttimes resorted thither with his disciples.)

5. Then saith he unto them, My soul is exceeding sorrowful, even unto death; tarry ye here, and watch with me.

6. And he went little farther and fell on his face and prayed, saying, O my Father-Mother, if it be possible, let this cup pass from me; nevertheless not as I will, but as Thou wilt.

7. And there appeared an angel unto him, from heaven strengthening him. And he cometh unto the disciples and finding them asleep, saith unto Peter, What, could ye not watch with me one hour?

8. Watch and pray that ye enter not into temptation: the spirit indeed is willing, but the flesh is weak.

9. He went away again a second time and prayed, saying, O my Father-Mother, if this cup may not pass away from me, except I drink it, Thy will be done.

10. And being in an agony he prayed more earnestly: and his sweat was as it were great drops of blood falling to the ground.

11. And he came and found them asleep again, for their eyes were heavy.

12. And he left them and went away again and prayed a third time, saying, O my Father-Mother, not my will but Thine be done, in earth as it is in heaven.

13. Then cometh he unto his disciples and saith unto them, Sleep on now, and take your rest; behold, the hour is at hand, and the Son of man is betrayed into the hands of sinners. Rise, let us be going: behold, he is at hand that doth betray me.

LECTION 77. 2. - Here the Lord addresses Simon, not Peter. In the A. V. confusion has arisen owing to the same name being given to two Apostles, and Peter is made to reply. It does not seem likely that one who thrice betrayed the Lord should by him have been placed in the highest authority, as it subsequently appears that Peter was.

Lection 78

The Betrayal by Judas Iscariot

1. AND it came to pass while Iesus yet spake, behold there came a multitude, and Judas that was called Iscariot went before them. For Judas, having received a band of men and officers from the chief priests and Pharisees, came thither with lanterns and torches and weapons.

2. Iesus therefore, knowing all things that should a come upon him, went forth and said unto them, Whom seek ye? They answered him, Iesus of Nazareth. Iesus saith unto them, I am he.

3. As soon then as he had said unto them, I am he, they went backward and fell to the ground. And when they arose, then asked he them again, Whom seek ye? And they said, Iesus of Nazareth. And Iesus answered, I have told you, I am he; if therefore ye seek me let these go their way.

4. Now he that betrayeth him gave them a sign, saying, Whomsoever I shall kiss, that same is he: hold him fast.

5. And forthwith he came to Iesus and said, Hail, Master; and kissed him. And Iesus said unto him. Friend, wherefore art thou come? Is it with a kiss that thou betrayest the Son of man?

6. Then Iesus said unto the chief priests and captains of the temple and the elders, which were come to him, Why ye come out as against a thief, with swords and staves? When I was daily with you in the temple, ye stretched forth no hands against me; but this is your hour, and the power of darkness.

7. Then came they and laid hands on Iesus. And Simon Peter stretched forth his hand, and drew his sword and struck a servant of the high priest's and smote off his ear.

8. Then said Iesus unto him, Put up again thy sword into its place; all they that take the sword shall perish by the sword. And Iesus touched his ear and healed him.

9. And he said unto Peter, Thinkest thou that I cannot now pray to my Parent, and He shall presently give me more than twelve legions of angels? But how then shall the scriptures be fulfilled, that thus it must be?

10. Then all the disciples forsook him and fled. And they that had laid hands on Iesus led him away to Caiaphas, the high

priest. But they brought him to Annas first because he was father-in-law to Caiaphas, who was the high priest for that same year.

11. Now Caiaphas was he who gave council to the Jews that it was expedient that one man should die for the sins of the people.

12. And the scribes and the elders were assembled together, but Peter and John and Simon and Jude followed far off unto the high priest's palace, and they went in and sat with the servants to see the end.

13. And they had kindled a fire in the midst of the hall, and when they were set down together, Peter sat down among them and warmed himself, and Simon also sat by him.

14. But a certain maid beheld him as he sat by the fire, and earnestly looked upon him and said, This man was also with him. And he denied him, saying, Woman, I know him not.

15. And after a little while, another saw him and said, Thou art also of them. And Simon said, Man, I am not.

16. And about the space of one hour another confidently affirmed, saying, Of a truth this fellow was with Iesus of Nazareth for his speech betrayeth him.

17. And Simon denied the third time with an oath, saying, I know not the man. And immediately, while he yet spake, the cock crew.

18. And the Lord turned and looked upon Simon. And Simon remembered the word of the Lord, how he had said unto him, Before the cock crow this day thou shalt deny me thrice. And Simon went out and wept bitterly.

LECTION 78. 12-18. -The belief that Peter denied his Master is probably owing to two of the Apostles bearing the same name, Simon Peter and Simon the Canaanite. Here we are given the right version. The error is one that might have been easily made. It is worthy of notice, that this ancient Gospel attributes to Simon (not to Simon Peter) the thrice denial of Iesus, and his fully exonerates Peter from the baseness generally attributed to him, and to which there is no allusion in his writings, but rather the reverse in the accepted gospel, where he was first to draw the sword in defense of his Master.

Lection 79

The Hebrew Trial Before Caiaphas.

1. THE high priest then asked Iesus of his disciples and of his doctrine, saying, How old art thou? Art thou he that said that our father Abraham saw thy day?

2. And Jesus answered, Verily before Abraham was I am. And the high priest said, Thou are not yet fifty years old. How sayest thou that thou hast seen Abraham? Who art thou? Whom makest thou thyself to be? What dost thou teach?

3. And Iesus answered him, I spake openly to the world; I even taught in the synagogue and in the temple, whither the Jews always resort; and in secret have I said nothing. Why asketh thou me? Ask them which heard me, what I have said unto them; behold, they know what I said.

4. And when he had thus spoken, one of the officers which stood by, struck Jesus with the palm of his hand, saying. Answerest thou the high priest so? Iesus answered him, If I have spoken evil, bear witness of the evil, but if well why smitest thou me?

5. Now the chief priests and elders, and all the council sought false witnesses against Iesus to put him to death; but found none; yea, many false witnesses came, yet they agreed not together.

6. At the last came two false witnesses. And one of them said, This fellow said, I am able to destroy the temple of God and to build it in three days. And the other said, This man said I will destroy this temple and build up another.

7. And the high priests arose and said unto him, Answerest thou nothing? What is it which these witnesses speak against thee? But Iesus held his peace. Now it was unlawful among the Hebrews to try a man by night.

8. And they said unto him, Art thou the Christ? tell us. And he said unto them, If I tell you, ye will not believe; and if I also ask you, ye will not answer me, nor let me go.

9. And they asked him further saying, Dost thou abolish the sacrifices of the law, and the eating of flesh as Moses commanded? And he answered, Behold, a greater than Moses is here.

10. And the high priest answered and said unto him, I adjure thee by the living God, that thou tell us whether thou be the Christ, the Son of God. Jesus saith unto him, thou hast said; and I say unto you, Hereafter shall ye see the Son of man sitting on the right hand of power and coming in the clouds of Heaven.

11. Then the high priest rent his clothes, saying, He hath spoken blasphemy; what further need have we of witnesses? Behold, now ye have heard his blasphemy. What think ye? They answered and said, He is worthy of death.

12. Then did they spit in his face and buffeted him; and others smote him with the palms of their hands, saying, Prophesy unto us, thou Christ, Who is he that smote thee?

13. Now when morning was come all the chief priests and the elders of the people, even the whole council held a consultation, and took council against Iesus to put him to death.

14. And they gave forth their sentence against Iesus, that he was worthy of death, and that he should be bound and carried away, and delivered unto Pilate.

LECTION 79. 2. -In a preceding Lection (LII.) the Jews at that time adjudged him then to be forty-five, and here Caiaphas, who must certainly have known his age, declared him to be "not yet 50," *ie.* about 49. This is borne out by the A. V. and by the testimony of S. Irenaeus, A.D. 120-22, and the testimony of S. Iohn the Apostle and his immediate disciples.

Lection 80

The Penance Of Judas.

1. NOW Judas, who had betrayed him, when he saw that he was condemned, repented himself, and brought again the thirty pieces of silver to the chief priests and elders, saying, I have sinned in that I have betrayed the innocent blood.

2. And they said, What is that to us? See thou to that. And he cast down the pieces of silver in the temple and departed and went out and hanged himself.

3. And the chief priests took the pieces of silver and said, It is not lawful for to put them into the treasury, because it is the price of blood.

4. And they took council and bought with them the potter's field, to bury strangers in. Wherefore that field was called Aceldama, that is, the field of blood, unto this day.

5. Then was fulfilled that which was spoken by Zachariah, the prophet, saying, They weighed for my price thirty pieces of silver. And they took the thirty pieces of silver, the price of him that was valued, whom they of the children of Israel did value, and gave them for the potteries field, and cast them to the potter in the House of the Lord.

6. Now, Iesus had said to his disciples, Woe unto the man who receiveth the mysteries, and falleth into sin thereafter.

7. For such there is no place of repentance in this cycle, seeing they have crucified afresh the Divine Offspring of God and man, and put the Christ within them to an open shame.

8. Such are worse than the beasts, whom ye ignorantly affirm to perish, for in your Scriptures it is written, That which befalleth the beast befalleth the sons of men.

9. All live by one breath, as the one dieth so dieth the other, so that a man hath no preeminence over a beast, for all go to the same place—all come from the dust and return to the dust together.

10. These things spake Iesus concerning them which were not regenerate, not having received the Spirit of Divine Love, who, once having received the Light, crucified the Son of God afresh, putting him to an open shame.

LECTION 80. 1. -The heading of this Lection in the A. V. is most misleading. "Penance," implying reparation of some kind (even though not of the right kind), is the more correct description of the act.

The Gospel of the Holy Twelve

Translated from the original Aramaic
by Rev. G.J.R. Ouseley

Section 9, Lections 81 Thru 89

Lection 81

The Roman Trial Before Pilate

1. THEN led they Jesus from Caiaphas unto the hall of judgment, to Pontius Pilate, the Governor, and it was early, and they themselves went not into the judgment hall, lest they should be defiled; but that they might keep the feast.

2. Pilate therefore went out unto them and said, What accusation bring ye against this man? They answered and said unto him, If he were not a malefactor, we would not have delivered him up unto thee. We have a law and by our law he ought to die, because he would change the customs and rites which Moses delivered unto us, yea, he made himself the Son of God.

3. Then said Pilate unto them, Take ye him, and Judge him according to your law. For he knew that for envy they had delivered him.

4. The Jews therefore said unto him, It is not lawful for us to put any man to death. So the saying of Iesus was fulfilled, which he spake, signifying what death he should die.

5. And they further accused him saying, We found this fellow perverting the nation, and forbidding to give tribute to Caesar, saying that he himself is Christ a King.

6. Then Pilate entered into the judgment hall again and called Iesus and said unto him, Art thou the King of the Jews? Iesus answered him, Sayest thou this thing of thyself, or did others tell it thee of me?

7. Pilate answered, Am I a Jew ? Thine own nation and the chief priests have delivered thee unto me; what hast thou done? Iesus answered, My kingdom is not of this world, if my kingdom were of this world, then would my servants fight, that I should not be delivered to the Jews; but now is my kingdom not from hence.

8. Pilate therefore said unto him, Art thou a King then? Iesus answered, Thou sayest that I am, yea, a King I am. To this end was I born and for this cause came I unto the world, that I should bear witness unto the truth. Every one that is of the truth heareth my voice.

9. Pilate said unto him, What is truth? Jesus said, Truth is from heaven. Pilate said, Then truth is not on earth. Jesus said unto Pilate, Believe thou, that truth is on earth amongst those who receive and obey it. They are of the truth who judge righteously.

10. And when he had heard this, he went out again unto the Jews and saith unto them, I find in him no fault at all. And when he was accused of the chief priests and elders he answered them nothing.

11. Then said Pilate unto him, Hearest thou not, how many things they witness against thee?

12. And he answered him never a word, insomuch that the governor marveled greatly, and again he said unto them, I find no fault in this man.

13. And they waxed the more fierce saying, He stirreth up the people, teaching throughout all Jewry, beginning from Galilee to this place. When Pilate heard of Galilee he asked, whether the man were a Galilean.

14. AND as soon as he knew that he belonged unto Herod's jurisdiction, he sent him to Herod, who himself also, was at Jerusalem at the time.

15. And when Herod saw Iesus he was exceedingly glad, for he was desirous to see him of a long season, because he had heard many things of him, and he hoped to have seen some miracle done by him.

16. Then he questioned with him in many words, but he answered him nothing. And the chief priests and scribes stood and vehemently accused him, and many false witnesses rose up against him, and laid to his charge things that he knew not.

17. And Herod with his men of war set him at nought, and mocked him, and arrayed him in a gorgeous robe and sent him again to Pilate. And the same day Pilate and Herod were made friends together, for before they were at enmity between themselves.

18. And Pilate went again into the Judgment Hall and saith unto Jesus, Whence art thou? But Iesus gave him no answer. Then saith Pilate unto him, Speakest thou not unto me? knowest thou not that I have power to crucify thee, and have power to release thee?

19. Iesus answered, Thou couldest have no power at all against me, except it were given thee from above, therefore he that delivered me unto thee hath the greater sin.

20. And from thenceforth Pilate sought to release him; but the Jews cried out, saying, If thou let this man go thou art no Caesar's friend, whosoever maketh himself a king speaketh against Caesar.

21. And Pilate called together the chief priests and rulers of the people. When he was set down on the judgement seat his wife sent unto him, saying. Have thou nothing to do with that just man, for I have suffered many things this day in a dream, because of him.

22. And Pilate said unto them, Ye have brought this man unto me, as one that perverteth the people, and behold I have

examined him before you, and have found no fault in this man touching those things: whereof ye accuse him. No, nor yet Herod, for I sent you to him, and lo nothing worthy of death was found in him.

23. But ye have a custom that I should release unto you one at the Passover, will ye therefore that I release unto you the King of the Jews?

24. Then cried they all again, saying, Not this man, but Barabbas. Now Barabbas was a robber. And, for sedition made in the city, and for murder, was cast into prison.

25. Pilate therefore, willing to release Iesus, spake again to them. Whether of the twain will ye that I release unto you; Iesus Barabbas, or Iesus which is called the Christ? They said, Barabbas

26. Pilate said unto them, What then shall I do with Iesus which is called the Christ? They all say unto him, Let him be crucified.

27. And the Governor said, Why what evil hath he done? But they cried out all the more saying, Crucify him, crucify him.

28. Pilate therefore went forth again and said unto him, Behold, again, I bring him forth to you, that ye may know that I find no fault in him, and again they cried out, Crucify him, crucify him.

29. And Pilate said unto them, the third time, Why, what evil hath he done? I have found no cause of death in him: I will therefore chastise him, and let him go.

30. And they were instant with loud voices, requiring that he might be crucified. And the voices of them and of the chief priests prevailed.

31. When Pilate saw that he could prevail nothing, but that rather a tumult was made, he took water, and washed his hands before the multitude, saying, I am innocent of the blood of this just person: see ye to it.

32. Then answered all the people, and said, His blood be on us and on our children. And Pilate gave sentence that it should be as they required. And he delivered Iesus to their will.

LECTION 81. 2. -This verse, suppressed by corruption of the Gospel, doubtless refers to the keeping the Passover within the gates without the slaying of So lamb, a capital offence by the law *(See* " New Aspects of Religion," by Dr. H. Pratt), or it might refer to keeping Passover the day before. There is much uncertainty on this point, the Gospels in the A. V. setting forth two different views, mutually contradicting each other, but neither of them implying necessarily the eating of a lamb by Jesus and his Apostles.

LECTION 81. 9. -These words, or the substance of them, are also to be found in one of the gnostic Gospels, which record many genuine sayings of the Master.

Lection 82

The Crucifixion

1. THEN released he Barabbas unto them, and when he had scourged Iesus he delivered him to be crucified. Then the soldiers of the governor took Iesus to the common hall and gathered unto him the whole band of soldiers.

2. And they stripped him and put on him a purple robe. And when they had plaited a crown of thorns they put it upon his head and a reed in his right hand, and they bowed the knee before him and mocked him, saying, Hail, King of the Jews!

3. Then came Iesus forth, wearing the crown of thorns, and the purple robe. And Pilate saith unto them, Behold the man!

4. When the chief priests therefore and officers saw him, they cried out, saying, Crucify him, crucify him. And Pilate saith unto them, Take ye him and crucify him, for I find no fault in him.

5. And they spit upon him, and took the reed and smote him on the head. And after that they had mocked him they took the robe off from him, and put his own raiment on him, and led him away to crucify him.

6. And as they led him away, they laid hold upon one Simon, a Cyrenian, coming out of the county, and on him they laid the cross that he might bear it after Iesus. And there followed him a great company of people and of women, which also bewailed and lamented him.

7. But Jesus, turning unto them, said, Daughters of Jerusalem, weep not for me, but weep for yourselves and for your children. For behold the days are coming in which they shall say, Blessed are the barren, and the wombs that never bare, and the paps which never gave suck.

8. Then shall they begin to say to the mountains, Fall on us; and to the hills, Cover us. For it they do these things in a green tree, what shall be done in the dry.

9. And there were also two other malefactors led with him to be put to death. And when they were come unto a place called Calvary, and Golgotha, that is to say a place of a skull, there they crucified him; and the malefactors, one on the right hand, and other on the left.

10. And it was the third hour when they crucified him, and they gave him vinegar to drink mingled with gall, and when he had tasted thereof, he would not drink. And Jesus said, Abba Amma, forgive them, for they know not what they do.

11. Then the soldiers, when they had crucified Jesus, took his raiment and made four parts, to every soldier a part; and also his vesture. Now the vesture was without seam, woven from the top throughout. They said therefore among themselves, Let us not rend it, but cast lots for it, whose it shall be.

12. That the scripture might be fulfilled, which saith, They parted my raiment among them, and for my vesture they did cast lots. These things therefore the soldiers did. And sitting down they watched him there.

13. And a superscription was also written over him in letters of Greek, and Latin, and Hebrew, This is the King of the Jews.

14. This title then read many of the Jews, for the place where Iesus was crucified was nigh to the city, and it was written in Hebrew and Greek and Latin. then said the chief priests of the Jews to Pilate, Write not, The King of the Jews, but that, he said, I am the King of the Jews. Pilate answered, What I have written, I have written.

15. And one of the malefactors which were hanged railed on him, saying, If thou be the Christ, save thy self and us. But the other answering rebuked him, saying, Dost not thou fear God, seeing thou art in the same condemnation? And we indeed justly, for we receive the due reward of our deeds, but this man hath done nothing amiss.

16. And he said unto Iesus, Lord remember me when thou comest into thy kingdom. And Iesus said unto him, Verily I say unto thee, to day shalt thou be with me in Paradise.

17. And they that passed by reviled him, wagging heir heads and saying, Thou that wouldst destroy the temple, and build it in three days, save thyself. If thou be the Son of God, come down from the Cross.

18. Likewise also the chief priests mocking him, while the scribes and elders said, He saved a lamb, himself he cannot save. If he be the King of Israel, let him now come down from the cross and we will believe him. He trusted in God, let Him deliver him now, if He will have him, for he said, I am the Son of God.

19. The usurers and the dealers in beasts and birds also cast the like things into his teeth, saying, Thou who drivest from the temple the traders in oxen and sheep and doves, art thyself but a sheep that is sacrificed.

20. Now from the Sixth hour there was darkness over all the land unto the Ninth hour, and some standing around, lighted their torches, for the darkness was very great. And about the Sixth hour Jesus cried with a loud voice, Eli, Eli, lame sabachthani? that, is to say, My God, My God, why hast Thou forsaken me ?

21. Some of them that stood there, when they heard that, said, This man calleth for Elias; others said, He calleth on the Sun. The rest said, Let be, let us see whether Elias will come to save him.

22. Now there stood by the cross of Iesus his mother and his mother's sister, Mary, the wife of Cleophas, and Mary Magdalene.

23. When Iesus therefore saw his mother, and the disciple standing by whom he loved, he saith unto his mother, Woman, behold thy son! And he said to the disciple, Behold thy mother! And from that hour that disciple took her into his own home.

24. After this, Iesus knowing that all things were now accomplished, that the scripture might be fulfilled, saith, I am athirst. And from a vessel they filled a sponge with vinegar and put it upon hyssop and put it to his mouth.

25. And Iesus cried with a loud voice, saying, Abba Amma, into Thy hand I commend my spirit.

26. When Iesus had therefore received the vinegar, he cried aloud, It is finished; and he bowed his head and gave up the ghost. And it was the ninth hour.

27. And behold there was great thunder and lightning, and the partition wall of the Holy place, from which hung the veil, fell down, and was rent in twain, and the earth did quake, and the rocks also were rent.

28. Now when the centurion and they that were with him watching Iesus, saw the earthquake and those things that were done, they feared greatly, saying, Truly this was a Son of God.

29. And many women were there, which followed from Galilee, ministering unto them, and among them were Mary the mother of James and Joses, and the mother of Zebedee's children and they lamented, saying, The light of the world is hid from our eyes, the Lord our Love is crucified.

30. Then the Jews, because it was the preparation, that the bodies should not remain upon the cross on the Sabbath, for that was a Paschal Sabbath, besought Pilate that their legs might be broken, and that they might be taken away.

31. Then came the soldiers, and brake the legs of the two who were crucified with him. But when they came to Jesus, and saw that he was dead already, they brake not his legs, but one of the soldiers with a spear pierced his heart and forthwith came there out blood and water.

32. And he that saw it bare record and his record is true, and he knoweth that he saith true, that ye might believe. For these things were done that the Scriptures might be fulfilled—A bone of him shall not be broken, and again—In the midst of the week the Messiah shall be cut off.

LECTION 82. 10-12. -Eli Reclus, a French writer, has some interesting remarks on the rite of human sacrifice as practised among the Khonds from time immemorial. The coincidences in the details are very striking, shewing the similarity of superstitious ideas in all countries and tribes of the primitive world-ideas which survive even in our own times " civilized ., as we boast them to be, but in reality savages when the skin deep "civilization" and culture are suddenly brushed away by some violent popular outburst, as in England, against the peaceful peoples of the Transvaal by which she brought herself to the lowest depths of infamy, and unwittingly clothed herself in the colour symbolic of dirt and mire.

v. 20. -In the Gospel attributed to Peter there is mention of the same circumstance. And to bring to mind, by symbolical art, this awful scene, among other reasons, the dark unbleached candles are lighted on the Altar on the day and at the hour when the Church commemorates the crucifixion of the Redeemer by an ingrate priesthood and people, when the light of the sun is shut out or obscured, and the chancels are draped in black.

v. 30. -It should be observed that in this Gospel, the mystically central organ of the Sacred Body, the "Heart" is emphasized rather than "his side," as in the A. V. on which last reading the strange custom of having a side entrance or porch to Churches is alleged to have been founded. The traditonal but corrupt reading of Gen. vi.16 has doubtless originated the error. *(See* "Original Genesis.")

v. 31. -They pierced his Sacred Heart with a spear, and this is symbolised in Christian Churches (which are generally cruciform either externally or internally where they are not circular), where the choir *(Cor.)* is in the intersection of nave and transept, and the altar of incense is *(ought to be)* in the midst under the great dome, symbolizing that the Sacred Heart of the Crucified is venerated from the centre to the extreme limits of Christendom-the Heart of God which embraces all creatures in its boundless love.

Lection 83

The Burial Of Jesus

1. NOW, when the even was come, Joseph of Arimathea, an honourable councillor, who also waited for the Kingdom of God, came and went in boldly unto Pilate and craved the body of Iesus. (He was a good man and just, and had not consented to the council and deed of them).

2. And Pilate marvelled if he were already dead, and calling unto him the centurion, he asked him whether he had been any while dead. And when he knew it of the centurion, he gave the body to Joseph. He came therefore, and took the body of Jesus.

3. And there came also Nicodemus, who at the first came to Iesus by night, and brought a mixture of myrrh and aloes, about an hundred weight. Then took they the body of Iesus and wound it in linen clothes with the spices, as the manner of the Jews is to bury.

4. Now in the place where he was crucified there was a garden, and in the garden a new sepulchre, wherein was never man yet laid. There laid they Iesus therefore, and it was about the beginning of the second watch when they buried him, because of the Jews' preparation day, for the sepulchre was nigh at hand.

5. And Mary Magdalene and the other Mary, and Mary the mother of Joses beheld where he was laid. There at the tomb they kept watch for three days and three nights.

6. And the women also, who came with him from Galilee, followed after, bearing lamps in their hands and beheld the sepulchre and how his body was laid, and they made lamentation over him.

7. And they returned and rested the next clay, being a high day, and on the day following they bought and prepared spices and ointments and waited for the end of the Sabbath.

8. Now the next day that followed, the chief priests and Pharisees came together unto Pilate, saying, Sir we remember that deceiver said, while he was yet alive, After three days I will rise again.

9. Command therefore that the sepulchre be made sure until the third day be past, lest his disciples come by night and steal him away, and say unto the people, He is risen from the dead, so the last error shall be worse than the first.

10. Pilate said unto them, Ye have a watch, go your way, make it as sure as you can. So they went and made the sepulchre sure, sealing the stone and setting a watch till the third day should be past.

LECTION 83. 5. -It has been maintained by some with no small degree of reason and probability, that the day of Crucifixion was not Friday, the day now observed by Christendom, but Wednesday (mid-week), by which date alone would be truly fulfilled the prophecy of Daniel, and the only sign of the truth of His mission which he would give to his generation. There shall no sign be given it, but the sign of the prophet Ionas, for as Ionas was three days and three nights in the whale's belly, so shall the Son of man be three days and three nights In the heart of the earth." Against this plain testimony there is of course the canonical record as we now have it including the frequent explanatory notes which may have been incorporated in very early times from the margin into the text, or interpolated in ignorance of the original script, which no man living has ever Been from this to the 10th century when all manuscripts were in the hands of the religious orders of the Church, and from them proceeded. If these words of Jesus be a genuine portion of the Gospel, as all admit they are, those notes of time, in the present accepted Gospels must be spurious, or the work of scribes who sought with honest and pious intent to harmonise the words of Scripture with the existent beliefs and observances of their age. In the gospel as now given there is absolutely nothing to militate against either of these views except the words of Jesus above cited, which cast the weight in favour of this chronological arrangement which interferes with nothing of Christian doctrine. Sunday, as now the day of his public entry into Ierusalem, preceded by the last anointing by Mary Magdalene on the eve before it. Monday, the day of evil counsel. Tuesday, the day of the Pascal feast of Christ. Wednesday, the day of the crucifixion, if not of the actual Jewish Passover. Thursday, Friday, Sabbath days of watch, of mourning and vigil. Sunday the day of the Resurrection, midnight or 8 a.m., early dawn " (after three days and three nights were fulfilled) and of the rising of many who slept and of their appearance in the holy City."

Lection 84

The Resurrection Of Jesus

1. NOW after the Sabbath was ended and it began to dawn, on the first day of the week, came Mary Magdalene to the sepulchre, bearing the spices which she had prepared, and there were others with her.

2. And as they were going, they said among themselves, who shall roll away the stone from the door of the sepulchre? For it was great. And when they came to the place and looked, they saw that the stone was rolled away.

3. For behold there was a great earthquake; and the angel of the Lord descended from heaven, and rolled back the stone from the door, and sat upon it. His countenance was like lighting and his raiment white as snow: And for fear of him the keepers did shake and became as dead men.

4. And the angel answered and said unto the women, Fear not ye, for I know that ye seek Iesus, which was crucified. He is not here: for he is risen, as he said.

5. Come, see the place where the Lord lay. And go quickly and tell his disciples that he is risen from the dead; and, behold he goeth before you into Galilee; there shall ye see him; lo, I have told you.

6. And they entered in and found not the body of Jesus. Then she ran and came to Simon Peter and the other disciple whom Jesus loved, and said unto them, They have taken away the Lord out of the sepulchre, and we know not where they have laid him.

7. And they ran and came to the scpulchre, and looking in, they saw the linen clothes lying, and the napkin that had been about his head not lying with the linen clothes, but wrapped up in a place by itself.

8. And it came to pass as they were much perplexed, behold, two angels stood by them in glistening garments of white, and said unto them, Why seek ye the living among the dead? He is not here, he is risen, and, behold, he goeth before you into Galilee, there shall we see him.

9. Remember ye not how he spake unto you, when he was yet in Galilee, that the Son of Man should be crucified and that he would rise again after the third day? And they remembered his words. And they went out quickly and fled from the sepulchre, for they trembled with amazement, and they were afraid.

10. NOW at the time of the earthquake, the graves were opened; and many of the saints which slept arose, and came out of the graves after his resurrection, and went into the city and appeared unto many.

11. But Mary stood without at the sepulchre weeping, and as she wept she again stooped down, and looked into the sepulchre and saw two angels in white garments, the one at the head, and the other at the feet, where the body of Iesus had lain. And they said unto her, Woman, why weepest thou?

12. She saith unto them, Because they have taken away my Lord, and I know not where they have laid him. And when she had thus said, she turned herself back, and saw Iesus standing, and knew not that it was Iesus.

13. Iesus saith unto her, Woman, why weepest thou? Whom seekest thou? She, supposing him to be the gardener, saith unto him, Sir, if thou have borne him hence, tell me where thou hast laid him, and I will take him away. Iesus said unto her, Mary, She turned herself and saith unto him, Rabboni; which is to say, Master.

14. Iesus saith unto her, Touch me not, for I am not yet ascended to my Father One with my Mother, but go to my brethren, and say unto them, I ascend unto my Parent and your Parent; to my God and your God.

15. And Mary Magdalene came and told the disciples that she had seen the Lord, and that he had spoken these things unto her, and commanded her to announce his resurrection from the dead.

LECTION 84. 6. - 0n this passage the celebrated writer M. Renan, bases his assertion that "but for Mary Magdalene Christianity would never have existed." It was she who first proclaimed the central fact -the Resurrection of the Lord. There is a true and a false side to his words.

Lection 85

Jesus Risen Again Appears To Two At Emmaus

1. AND behold, two of them went that same day to a village called Emmaus, which was from Jerusalem about threescore furlongs. And they talked together of all these things which had happened.

2. And it came to pass, that, while they communed together and reasoned, Iesus himself drew near, and went with them. But their eyes were holden that they should not know him.

3. And he said unto them, What manner of communications are these that ye have one with another, as ye walk and are sad?

4. And the one of them, whose name was Cleophas, answering, said unto him, Art thou only a stranger in Jerusalem and hast not known the things which are come to pass there in these days? And he said unto them, What things?

5. And they said unto him, Concerning Iesus of Nazareth who was a Prophet mighty in deed and word before God and all the people; and how the chief priests and our rulers delivered him to be condemned to death, and have crucified him. But we trusted that it had been he which should have redeemed Israel; and beside all this three days have passed since these things were done.

6. Yea, and certain women also of our company made us astonished, which were early at the sepulchre; and when they found not his body, they came saying, that they had also seen a vision of angels, who said that he was alive.

7. And certain of them who were with us went to the sepulchre, and found it even so as the women had said; but him they saw not.

8. Then he said unto them, O fools and slow of heart to believe all that the prophets have spoken; Ought not Christ to have suffered these things, and then to enter into his glory?

9. And beginning at Moses and all the prophets, he expounded unto them in all the scriptures, the things concerning himself.

10. And they drew nigh unto the village whither they went; and he made as though he would have gone further. But they constrained him, saying, Abide, with us, for it is toward evening, and the day is far spent. And he went in to tarry with them.

11. And it came to pass as he sat at table with them, he took bread and the fruit of the vine, and gave thanks, blessed, and brake, and gave to them. And their eyes were opened, and they knew him; and he vanished out of their sight.

12. And they said one to another, Did not our hearts burn within us while he talked with us by the way, and while he opened to us the scriptures? And they rose up the same hour and returned to Jerusalem, and found the twelve gathered together, and them that were with them, saying, The Lord is risen indeed, and hath appeared to Simon.

13. And they told what things were done in the way and how he was known of them in breaking of bread.

14. Now while they had been going to Emmaus, some of the watch came into the city, and showed unto Caiaphas what things had been done.

15. And they assembled with the elders and took council and said, Behold, while the soldiers slept, some of his disciples came and took his body away; and is not Joseph of Arimathea one if his disciples?

16. For this cause then did he beg the body from Pilate that he might bury it in his garden in his own tomb. Let us therefore give money to the soldiers, saying, say ye, His disciples came by night and stole him away while we slept. And if this come to the ears of the governor we will persuade him, and secure you.

LECTION 85. 15. -These words though not fully given in the A. V. have been made the basis of an attempted explanation by M. Renan, who could not receive the alternative view that the body of Jesus was dematerialized, rose, and then appeared in spiritual form, which view is held by believers of modern manifestations

Lection 86

Iesus Appears In The Temple and Blood Sacrifices Cease

1. THE same day, at the time of sacrifice in the Temple there appeared among the dealers in beasts and in birds, One clothed in white raiment, bright as light, and in his hand a whip of seven cords.

2. And at the sight of him, those who sold and bought fled in terror, and some of them fell as dead men, for they remembered how before his death Iesus had driven them away from the Temple enclosure, in like manner.

3. And some declared that they had seen a spirit. And others that they had seen him who was crucified and that he had risen from the dead.

4. And the sacrifices ceased that day in the Temple, for all were in fear, and none could be had to sell or to buy, but, rather, they let their captives go free.

5. And the priests and elders caused a report to be spread, That they who had seen it were drunken, and had seen nothing. But many affirmed that they had seen him with their own eyes, and felt on their backs the scourge, but were powerless to resist, for when some of the bolder among them put forth their hands, they could not seize the form which they beheld, nor grasp the whip which chastised them.

6. And from that time, these believed in Iesus, that he was sent from God, to deliver the oppressed, and free those that were bound. And they turned from their ways and sinned no longer.

7. To others he also appeared in love and mercy and healed them by his touch, and delivered them from the hands of the persecutor. And many like things were reported of him, and many said, Of a truth the Kingdom is come.

8. And some of those who had slept and risen, when Iesus rose from the dead appeared, and were seen by many in the holy City, and great fear fell upon the wicked, but light and gladness came to the righteous in heart.

LECTION 86. 4. -"The sacrifices ceased that day" -here is not meant for any permanence (as generally believed) for they went on, we are told, for forty years, till the destruction of the Temple by the Romans.

Lection 87

Jesus Appeareth To His Disciples.

1. THEN the same day at evening, being the first day of the week, when the doors were shut where the disciples were assembled for fear of the Jews, came Jesus and stood in the midst, and saith unto them, Peace be unto you. But they were affrighted and supposed that they had seen a spirit.

2. And he said unto them, Behold, it is I myself, like as ye have seen me aforetime. A spirit can in deed appear in flesh and bones as ye see me have. Behold my hands and my feet, handle and see.

3. And when he had so said, he shewed unto them his hands and his Heart. Then were the disciples glad, when they saw the Lord.

4. For Thomas, called Didymus, one of the disciples, had said unto them, Except I shall see in his hands the print of the nails, and thrust my hand into his heart, I will not believe. Then saith he to Thomas, Behold my hands, my heart, and my feet; reach

hither thy hands, and be not faithless but believing.

5. And Thomas answered and said unto him, My Lord and my God! And Iesus saith unto him, Thomas, because thou hast seen me, thou hast believed; blessed are they that have not seen and yet have believed.

6. Then saith Iesus unto them again, Peace be unto you, as Abba Amma hath sent me, even so send I you. And when he had said this he breathed on them and said unto them, Receive ye the Holy Ghost; preach the Gospel, and anounce ye unto all nations; the resurrection of the Son of Man.

7. Teach ye the holy law of love which I have delivered unto you. And whosoever forsake their sins, they are remitted unto them, and whosoever continue in their sins they are retained unto them.

8. **Baptise them who believe and repent,** bless and anoint them, and offer ye the pure Oblation of the fruits of the earth, which I have appointed unto you for a Memorial of me.

9. Lo, I have given my body and my blood to be offered on the Cross, for the redemption of the world from the sin against love, and from the bloody sacrifices and feasts of the past.

10. And ye shall offer the Bread of life, and the Wine of salvation, for a pure Oblation with incense, as it is written of me, and ye shall eat and drink thereof for a memorial, that I have delivered all who believe in me from the ancient bondage of your ancestors.

11. **For they, making a god of their belly, sacrificed unto their god the innocent creatures of the earth, in place of the carnal nature within themselves.**

12. And eating of their flesh and drinking of their blood to their own destruction, corrupted their bodies and shortened their days, even as the Gentiles who knew not the truth, or who knowing it, have changed it into a lie.

13. As I send you, so send ye others also, to do these things in my Name, and he laid his hands upon them.

14. In the like manner as the Apostles, so also be ordained Prophets and Evangelists and Pastors, a Holy Priesthood, and afterwards he laid his hand upon those whom they chose for Deacons, one for each of the fourfold twelve.

15. And these are for the rule and guidance of the Church Universal, that all may be perfected in their places in the Unity of the Body of the Christ.

LECTION 87. 1. -The power to come in, or to go out through closed doors, has been shown in modern times to be no impossibility, but a proven fact in psychological phenomena. The words here do not *necessarily* imply that such manifestation took place. It is not said, "they were locked," but the power of the Spirit to materialize, and dematerialize, Bond appear in human form (under certain conditions) is too well known to be denied." Report of Dialectical Society on Spiritual Phenomena," etc.

LECTION 87. 2. -The contradiction in the A.V. is here no longer seen. That a spirit can *appear* in flesh and bones has been testified over by thousands of competent witnesses in this as well as other ages. There is no death, and the returning spirit can *appear* in any form. Of these things we are witnesses.

v. **8.**-A similar passage to this occurs in the "Pistis Sophia," an ancient gnostic Gospel.

Lection 88

The Eighth Day After The Resurrection

1. AND after seven days again, his disciples were within the Upper Room; then came Iesus, the doors being shut, and stood in their midst and said, Peace be unto you, and he was known unto them in the holy Memorial.

2. And he said unto them. Love ye one another and all the creatures of God. Yet I say unto you, not all are men, who are in the form of man. Are they men or women in the image of God whose ways are ways of violence, of oppression and wrong, who choose a lie rather than the truth?

3. **Nay, verily, till they are born again, and receive the Spirit of Love and Wisdom within their hearts. Then only are they sons and daughters of Israel, and being of Israel they are children of God, And for this cause came I into the world, and for this I have suffered at the hands of sinners.**

4. These are the words which I spake unto you, while I was yet with you, that all things must be fulfilled which were written in the law of Moses and in the prophets, and in the psalms, concerning me.

5. And Iesus said, I stood in the midst of the world, and in the flesh was I seen and heard, and I found all men glutted with their own pleasures, and drunk with their own follies, and none found I hungry or athirst for the wisdom which is of God. My soul grieveth over the sons and daughters of men because they are blind in their heart, and in their soul are they deaf and hear not my voice.

6. Then opened he their understanding, that they might understand the scriptures. And said unto them, Thus it is written, and thus it behooved the Christ to suffer, and to rise from the dead after the third day. And that repentance and remission of sins should be preached in my name among all nations, beginning at Jerusalem. And ye are witnesses of these things.

7. And, behold, I send the promice of my Parent upon you, even of my Father One with my Mother, Whom ye have not seen on the earth. For I say unto you of a truth, as the whole world have been ruined by the sin and vanity of woman, so by the simplicity and truth of woman shall it be saved, even by you shall it be saved.

8. Rejoice therefore and be ye glad, for ye are more blessed than all who are on earth, for it is ye, my twelve thousand who shall save the whole world.

9. Again I say unto you when the great tyrant and all the seven tyrants began to fight in vain against the Light, they knew not with Whom or What they fought.

10. For they saw nothing beyond a dazzling Light, and when they fought they expended their strength one against another, and so it is.

11. For this cause I took a fourth part of their strength, so that they might not have such power, and prevail in their evil deeds.

12. For by involution and evolution shall the salvation of all the world be accomplished: by the Descent of Spirit into matter, and the Ascent of matter into Spirit, through the ages.

LECTION 88. 5. -Most affecting is this, the experience of all who in this world of madness and unreason attempt to declare the whole counsel of God. It broke the heart of Jesus, it crushes the heart of every prophet or apostle worker for good, filled with his spirit- " Jerusalem, Jerusalem I would - but *ye would not.*"

Lection 89

Jesus Appeareth At The Sea Of Tiberias

1. AFTER these things Jesus shewed himself again to the disciples at the sea of Tiberias, and on this wise shewed he himself. There were together Simon, Peter, and Thomas, called Didymus, and Nathanael of Cana in Galilee, and James and John and two other of his disciples.

2. And Peter saith unto them, I go a fishing. They say unto him, We also go with thee. They went forth and entered into a ship immediately, and that night they caught nothing. And when the morning was now come, Jesus stood on the shore, but the disciples knew not that it was Jesus.

3. Then Jesus said unto them, Children, have ye any meat? They answered him, Nay, Lord, not enough for all; there is naught but a small loaf, a little oil, and a few dried fruits. And he said unto them, Let these suffice; come and dine.

4. And he blessed them, and they ate and were filled, and there was a pitcher of water also, and he blessed it likewise, and lo, it was the fruit of the vine.

5. And they marvelled, and said. It is the Lord. And none of the disciples dost ask him. Who art thou? knowing it was the Lord.

6. This is now the sixth time that Iesus shewed himself to his disciples, after that he was risen from the dead. So when they had dined, Iesus saith to Peter, son of Jonas, lovest thou me more than these? He saith unto him, Yea, Lord, thou knowest that I love thee. He saith unto him, Feed my lambs. He saith unto him again the second time, Peter, son of Jonas, lovest thou me? He saith unto him, Yea, Lord thou knowest that I love thee. He said unto him. Feed my sheep.

7. He saith unto him the third time, Peter, son of Jonas, lovest thou me? Peter was grieved because he said unto him the third time, Lovest thou me ? And he said unto him, Lord, thou knowest all things; thou knowest that I love thee.

8. Iesus saith unto him, Feed my Flock. Verily verily, I say unto thee, thou art a rock from the Rock, and on this rock will I build my Church, and I will raise thee above my twelve to be my vicegerent upon earth for a centre of Unity to the twelve, and another shall be called and chosen to fill thy place among the twelve, and thou shalt be the Servant of servants and shalt feed my rams, my sheep and my lambs.

9. And yet another shall arise and he shall teach many things which I have taught you already, and he shall spread the Gospel among the Gentiles with great zeal. But the keys of the Kingdom will I give to those who succeed thee in my Spirit and obeying my law.

10. And again I say unto thee. When thou wast young thou girdedst thyself and walketh whither thou wouldst, but when thou shalt be old, thou shalt stretch forth thy hands and another shall gird thee and carry thee whither thou wouldst not. This spake he, signifying by what death he should glorify God.

11. And when he had spoken this he saith unto him, Follow me. Then Peter, turning about, seeth the disciple whom Iesus loved following. Peter seeing him, saith to Iesus, Lord and what shall this man do? Iesus saith unto him, If I will that he tarry till I come, what is that to thee? follow thou me.

12. Then went this saying abroad among the brethren that disciple should not die: yet Iesus said not unto him, He shall not die, but, if I will that he tarry till I come, what is that to thee.

LECTION 89. 2. -"That night they caught nothing" -henceforth their labours were to be in the Spiritual Kingdom to save souls -not destroy them- by bringing them within the Church of Christianity, from barbarism and darkness to reason and light and love.

The Gospel of the Holy Twelve

Translated from the original Aramaic
by Rev. G.J.R. Ouseley

Section 10, Lections 90 Thru 96

Lection 90

What Is Truth?

1. AGAIN the twelve were gathered together in the Circle of palm trees, and one of them even Thomas said to the other, What is Truth? for the same things appear different to different minds, and even to the same mind at different times. What, then, is Truth?

2. And as they were speaking Jesus appeared in their midst and said, Truth, one and absolute, is in God alone, for no man, neither any body of men, knoweth that which God alone knoweth, who is the All in All.. To men is Truth revealed, according to their capacity to understand and receive.

3. The One Truth hath many sides, and one seeth one side only, another seeth another, and some see more than others, according as it is given to them.

4. Behold this crystal: how the one light its manifest in twelve faces, yea four times twelve, and each face reflecteth one ray of light, and one regardeth one face, and another another, but it is the one crystal and the one light that shineth in all.

5. Behold again, When one climbeth a mountain and attaining one height, he saith, This is the top of the mountain, let us reach it, and when they have reached that height, lo, they see another beyond it until they come to that height from which no other height is to be seen, if so be they can attain it.

6. So it is with Truth. I am the Truth and the Way and the Life, and have given to you the Truth I have received from above. And that which is seen and received by one, is not seen and received by another. That which appeareth true to some, seemeth not true to others. They who are in the valley see not as they who are on the hill top.

7. But to each, it is the Truth as the one mind seeth it, and for that time, till a higher Truth shall be revealed unto the same: and to the soul which receiveth higher light, shall be given more light. Wherefore condemn not others, that ye be not condemned.

8. As ye keep the holy Law of Love, which I have given unto you, so shall the Truth be revealed more and more unto you, and the Spirit of Truth which cometh from above shall guide you, albeit through many wanderings, into all Truth, even as the fiery cloud guided the children of Israel through the wilderness.

9. Be faithful to the light ye have, till a higher light is given to you. Seek more light, and ye shall have abundantly; rest not, till ye find.

10. God giveth you all Truth, as a ladder with many steps, for the salvation and perfection of the soul, and the truth which seemeth to day, ye will abandon for the higher truth of the morrow. Press ye unto Perfection.

11. Whoso keepeth the holy Law which I have given, the same shall save their souls, however differently they may see the truths which I have given.

12. Many shall say unto me, Lord, Lord, we have been zealous for thy Truth. But I shall say unto them, Nay, but, that others may see as ye see, and none other truth beside. Faith without charity is dead. Love is the fulfilling of the Law.

13. How shall faith in what they receive profit them that hold it in unrighteousness? They who have love have all things, and without love there is nothing worth. Let each hold what they see to be the truth in love, knowing that where love is not, truth is a dead letter and profiteth nothing.

14. There abide Goodness, and Truth, and Beauty, but the greatest of these is Goodness. If any have hatred to their fellows, and harden their hearts to the creatures of God's hands, how can they see Truth unto salvation, seeing their eyes are blinded and their hearts are hardened to God's creation?

15. As I have reveived the Truth, so have I given it to you. Let each receive it according to their light and ability to understand, and persecute not those who receive it after a different interpretation.

16. For Truth is the Might of God, and it shall prevail in the end over all errors. But the holy Law which I have given is plain for all, and just and good. Let all observe it for the salvation of their souls.

LECTION 90. 4. -The art of cutting and polishing glass and stone was well known in Phoenicia and Egypt, before the Christian era, and in Pompeii numbers of such crystals were found in great variety. It is a beautiful symbol appealing to the mind.

LECTION 90. 12. -Our Lord never damned or blamed those who could not see the divine truths, which he taught, and receive them. He had patience with them, as being without the fold, without light, and not admissible to the Kingdom, so long as they remained in their darkness and impenitence and self-doomed to eternal death if they persisted.

Lection 91

The Order of the Kingdom. (*Part I.*)

1. In that time after Iesus had risen from the dead he tarried ninety days with Mary his mother and Mary Magdalene, who anointed his body, and Mary Cleophas and the twelve, and their fellows, instructing them and answering questions concerning the kingdom of God.

2. And as they sat at supper—when it was even— Mary Magdalene asked him, saying, Master, wilt thou now declare unto us the Order of the Kingdom?

3. And Iesus answered and said, Verily I say unto thee, O Mary, and to each of any disciples, The kingdom of Heaven is within you. But the time cometh when that which is within shall be made manifest in the without, for the sake of the world.

4. Order indeed is good, and needful, but before all things is love. Love ye one another and all the creatures of God, and by this shall all men know that ye are my disciples.

5. AND one asked him saying, Master, wilt thou that infants be received into the congregation in like manner as Moses commanded by circumcision? And Jesus answered, For those who are in Christ there is no cutting of the flesh, nor shedding of blood.

6. Let the infant of eight clays be Presented unto the Father-Mother, who is in Heaven, with prayer and thanksgiving, and let a name be given to it by its parents, and let the presbyter sprinkle pure water upon it, according to that which is written in the prophets, and let its parents see to it that it is brought up in the ways of righteousness, neither eating flesh, nor drinking strong drink, nor hurting the creatures which God hath given into the hands of man to protect .

7. AGAIN one said unto him, Master, how wilt thou when they grow up? And Jesus said, After seven years, or when they begin to know the evil from the good, and learn to choose the good, let them come unto me and receive the blessing at the hands of the presbyter or the angel of the church with prayer and thanksgiving, and let them be admonished to keep from flesh eating and strong drink, and from hunting the innocent creatures of God, for shall they be lower than the horse or the sheep to whom these things are against nature?

8. And again he said, If there come to us any that eat flesh and drink strong drink, shall we receive them? And Iesus said unto him, Let such abide in the outer court till they cleanse themselves from these grosser evils; for till they perceive, and repent of these, **they are not fit to receive the higher mysteries.**

9. AND another asked him saying, When wilt thou that they receive Baptism? And Iesus answered, After another seven years, or when they know the doctrine, and do that which is good, and learn to work with their own hands, and choose a craft whereby they may live, and are stedfastly set on the right way. Then let them ask for initiation, and let the angel or presbyter of the church examine them and see if they are worthy, and let him offer thanksgiving and prayer, and bury them in the waters of separation, that they may rise to newness of life, confessing God as their Father and Mother, vowing to obey the Holy Law, and keep themselves separate from the evil in the world.

10. AND another asked him, Master, at what time shall they receive the Anointing? And Iesus answered, When they have reached the age of maturity, and manifested in themselves the sevenfold gifts of the Spirit, then let the angel offer prayer and thanksgiving and seal them with the seal of the Chrism. It is good that all be tried in each degree seven years. Nevertheless let it be unto each according to their growth in the love, and the wisdom of God.

LECTION 91. 5. -The idea of baptizing unconscious infants seems never to have entered the mind of Jesus. He blessed them, but he also blessed other animals, and things that had no sentient life. Baptism implies belief and confession of faith and repentance from evil works and ways.

LECTION 91. 6. -0ver 2,000 years before Christ there existed on the shores of Lake Meeris, in Egypt, a labyrinth of seven circular wall-enclosed winding paths, represented by Boticelli in one of his engravings, which we here reproduce adapted for Christian rites. This was used by the Egyptians in their initiations as a symbol of life, and the wanderings of the soul in the flesh, till "seven times seven" times purified and meet to appear before God. **Cont**

LECTION 91. 7-8. -In the Editor's former work "Palingenesia, or Earth's New Birth," 1884, incorporating some Ideas from this Gospel (part of which he had then received) these two rites referred to, by some oversight were transposed. Here, as in" Church of the Future" 1896, by the same Editor, the correct order is given. It is at present out of print.

Lection 92

The Order of the Kingdom. (*Part II.*)

1. AND another asked him saying, Master, wilt thou that there be marriages among us as it is among the nations of earth? And Iesus answered, saying, Among some it is the custom that one woman may marry several men, who shall say unto her, Be thou our wife and take away our reproach. Among others it is the custom, that one man may marry several women, and who shall say unto him, Be thou our husband and take away our reproach, for they who love feel it is a reproach to be unloved.

2. But unto you my disciples, I shew a better and more perfect way, even this, that marriage should be between one man and one woman, who by perfect love and sympathy are united, and that while love and life do last, howbeit in perfect freedom. But let them see to it that they have perfect health, and that they truly love each other in all purity, and not for worldly advantage only, and then let them plight their troth one to another before witnesses.

3. Then, when the time is come, let the angel or presbyter offer prayer and thanksgiving and bind them with the scarlet cord, if ye will, and crown them, and lead them thrice around the altar and let them eat of one bread and drink of one cup. Then holding their hands together, let him say to them in this wise, Be ye two in one, blessed be the holy union, you whom God doth join together let no man put asunder, so long as life and love do last.

4. And if they bear children, let them do so with discretion and prudence according to their ability to maintain them. Nevertheless to those who would be perfect and to whom it is given, I say, let them be as the angels of God in Heaven, who neither marry nor are given in marriage, nor have children, nor care for the morrow, but are free from bonds, even as I am, and keep and store up the power of God within, for their ministry, and for works of healing, even as I have done. But the many cannot receive this saying, only they to whom it is given.

5. AND another asked him saying, Master, in what manner shall we offer the Holy Oblation? And Iesus answered, saying, The oblation which God loveth in secret is a pure heart. But for a Memorial of worship offer ye unleavened bread, mingled wine, oil and incense. When ye come together in one place to offer the Holy Oblation, the lamps being lighted, let him who presideth, even the angel of the church, or the presbyter, having clean hands and a pure heart, take from the things offered, unleavened bread and mingled wine with incense.

6. And let him give thanks over them and bless them, calling upon the Father-Mother in Heaven to send their Holy Spirit that it may come upon and make them to be the Body and Blood, even the Substance and Life of the Eternal, which is ever being broken and shed for all.

7. And let him lift it up toward Heaven and pray for all, even for those who are gone before, for those who are yet alive, and for those who are yet to come As I have taught you, so pray ye, and after this let him break the bread and put a fragment in the cup, and then bless the holy union, and then let him give unto the faithful, saying after this manner, This is the body of the Christ even the substance of God (ever being broken and shed, for you and for all), unto eternal life. As ye have seen me do, so do ye also, in the spirit of love, for the words I speak unto you, they are spirit and they are life.

LECTION 92. 4. -Here we have further proof, if any were needed, that Jesus was brought up in the tenets and customs of the **Essenes**. See "Christianity and Buddhism" (a remarkable book by Arthur Lillie) for the full discussion of the subject.

v. **6.** -Similar were the rites of Mithra. From the days of Noah and Melchizedek these pure mysteries were celebrated -though not in the fulness of the light of Christ.

Lection 93

The Order of the Kingdom. (*Part III.*)

1. AND another spake, saying, Master, if one have committed a sin, can a man remit or retain his sin? And Iesus said, God forgiveth all sin to those who repent, but as ye sow, so also must ye reap; Neither God nor man can remit the sins of those who repent nor nor forsake their sins; nor yet retain the sins of those who forsake them. But if one being in the spirit seeth clearly that any repent and forsake their sins, such may truly say unto the penitent, Thy sins are forgiven thee, for All sin is remitted by repentance and amendment and they are loosed from it, who forsake it and bound to it, who continue it.

2. Nevertheless the fruits of the sin must continue for a season, for as we sew so must we reap, for God is not mocked, and they who sow to the flesh shall reap corruption, they who sow to the spirit shall reap life, everlasting. Wherefore if any forsake their sins and confess them, let the presbyter say unto such in this wise, May God forgive thee thy sins, and bring thee to everlasting life. All sin against God is forgiven by God, and sin against man by man.

3. AND another asked him, saying, If any be sick among us, shall we have power to heal even as thou dost? And Jesus answered, This power cometh of perfect chastity and of faith. They who are born of God keep their seed within them.

4. Nevertheless if any be sick among you, let them send for the presbyters of the church that they may anoint them with oil of olive in the Name of de Lord, and the prayer of faith, and the going out of power, with the voice of thanksgiving, shall raise them up, if they are not detained by sin, of this, or a former life.

5. AND another asked him saying, Master, how shall the holy assembly be ordered and who shall minister therein? And Jesus answered. When my disciples are gathered in my name let them choose from among themselves true and faithful men and women, who shall be ministers and counsellors in temporal things and provide for the necessities of the poor, and those who cannot work, and let these look to the ordering of the goods of the church, and assist at the Oblation, and let these be your deacons, with their helps.

6. And when these have given proof, of their ministry, let them choose from them, those who have spiritual gifts, whether of guidance, or of prophecy, or of preaching and of teaching and healing, that they may edify the flock, offer the holy Oblation and minister the mysteries of God and let these be your presbyter, and their helps.

7. And from these who have served well in their degree let one be chosen who is counted most worthy, and let him preside over all and he shall be your Angel. And let the Angel ordain the deacons and consecrate the presbyters—anoint them and laying their hands upon them and breathing upon them that they may receive the Holy Spirit for the office to which they are called. And as for the Angel let one of the

higher ministry anoint and consecrate him, even one of the Supreme Council.

8. For as I send Apostles and Prophets so also I send Evangelists And Pastors—the eight and forty pillars of the tabernacle—that by the ministry of the four I may build up and perfect my Church. and they shall sit in Jerusalem a holy congregation, each with his helper and deacon, and to them shall the scattered congregations refer in all matters pertaining to the Church. And as light cometh so shall they rule and guide and edify and teach my holy Church. They shall receive light from all, and to all shall they give more light.

9. And forget not with your prayers and supplications intercessions and giving of thanks, to offer the incense, as it is written in the last of your prophets, saying, From the rising of the sun unto the setting of the same incense shall be offered unto My Name in all places with a pure oblation, for My Name shall be great among the Gentiles.

10. For verily I say unto you, incense is the memorial of the intercession of the saints within the veil, with words that cannot be uttered.

Lection 94

The Order of the Kingdom. (*Part IV.*)

1. AND another asked him, saying, Master, how wilt thou that we bury our dead? And Iesus answered, Seek ye council of the deacons in this matter, for it concerneth the body only. Verily, I say, unto you there is no death to those who believe in the life to come. Death, as ye deemed it, is the door to life, and the grave is the gate to resurrection, for those who believe and obey. Mourn ye not, nor weep for them that have left you, but rather rejoice for their entrance into life.

2. As all creatures come forth from the unseen into this world, so they return to the unseen, and so will they come again till they be purified. Let the bodies of them that depart be committed to the elements, and the Father-Mother, who reneweth all things, shall give the angels charge over them, and let the presbyter pray that their bodies may rest in peace, and their souls awake to a joyful resurrection.

3. There is a resurrection from the body, and there is a resurrection in the body. There is a raising out of the life of the flesh, and there is a falling into the life of the flesh. Let prayer be made For those who are gone before, and For those that are alive, and For those that are yet to come, for all are One family in God. In God they live and move and have their being.

4. The body that ye lay in the grave, or that is consumed by fire, is not the body that shall be, but they who come shall receive other bodies, yet their own, and as they have sown in one life, so shall they reap in another. Blessed are they who have worked righteousness in this life, for they shall receive the crown of life.

5. AND another asked him, saying, Master, under the law Moses clad the priests with garments of beauty for their ministration in the Temple. Shall we also clothe them to whom we commit the ministry of sacred things as thou hast taught us? And Iesus answered, White linen is the righteousness of the Saints, but the time truly cometh when Zion shall be desolate, and after the time of her affliction is past, she shall arise and put on her beautiful garments as it is written.

6. But seek ye first the kingdom of righteousness, and all these things shall be added unto you. In all things seek simplicity, and give not occasion to vain glory. Seek ye first to be clothed with charity, and the garment of salvation and the robe of righteousness.

7. For what doth it profit if ye have not these? As the sound of brass and tinkling of cymbal are ye, if ye have not love. Seek ye righteousness and love and peace, and all things of beauty shall be added to you.

8. AND yet another asked him, saving, Master, how many of the rich and mighty will enter into life and join us who are poor and despised. How, then, shall we carry on the work of God in the regeneration of mankind? And Iesus said, This also is a matter for the deacons of the church in council with the elders.

9. But when my disciples are come together on the Sabbath, at even, or in the morning of the first day of the week, let them each bring an offering of a tithe, or the tithe of a tithe of their increase, as God doth prosper them, and put it in the treasury, for the maintenance of the church and the ministry, and the works thereof. For I say unto you, it is more blessed to give than to receive.

10. So shall all things be done, decently and in order, And the rest will the Spirit set in order who proceedeth from the Father-Mother in heaven. I have instructed you now in first principles, and, lo, I am with you always, even unto the end of the Age.

LECTION 94. 7. -From this, as from other words of the Master on previous occasions, it is evident that his servant Paul borrowed from him many of the ideas, and similes and wise sayings scattered through his Epistles, and not Paul only, but also the other Apostles. *(See* also verse 9).

v. 10. -It has been alleged that the laying down of rites and ordinances for Christianity has been the cause of division and strife in all countries. Nay, rather have not these divisions and dissensions been caused by the omission of the directions given by the One Head acknowledged by all during the period between his resurrection and ascension and the generation immediately after, and the handling of them down by that tradition so liable to corruption in place of the written record. But much more were these divisions and dissensions caused by the interpolation of dogmas not making for goodness and unity, by the suppression from the records of the vital essence in the holy law given by Iesus on the Mount, which, had it been preached and known and obeyed by all, would have made the earth a paradise in place of a hen for the weak and the helpless.

Lection 95

The Ascension.

1. AND Iesus after he had shewed himself alive to his disciples after his resurrection, and sojourned with them for ninety days, teaching and speaking of the Kingdom, and the things pertaining to the Kingdom of God, and had finished all things that he had to do, led forth the twelve with Mary Magdalene, and Joseph his father and Mary his mother, and the other holy women as far as Bethany to a mountain called Olivet, where he had appointed them.

2. And when they saw him as he stood in the midst of them, they worshipped him, but some doubted. And Iesus spake unto them, saying, Behold, I have chosen you from among men, and have given you the Law, and the Word of truth.

3. I have set you as the light of the world, and as a city that cannot be hid. **But the time cometh when darkness shall cover the earth, and gross darkness the people, and the enemies of truth and righteousness shall rule in my Name, and set up a kingdom of this world, and oppress the peoples, and cause the enemy to blaspheme, putting for my doctrines the opinions of men, and teaching in my Name that which I have not taught, and darkening much that I have taught by their traditions.**

4. **But be of good cheer, for the time will also come when the truth they have hidden shall be manifested, and the light shall shine, and the darkness shall pass away, and the true kingdom shall be established which shall be in the world, but not of it,** and the Word of righteousness and love shall go forth from the Centre, even the holy city of Mount Zion, and the Mount which is in the land of Egypt shall be known as an altar of witness unto the Lord.

5. And now I go to my Parent and your Parent, my God and your God. But ye, tarry in Jerusalem, and abide in prayer, and after seven days ye shall receive power from on high, and the promise of the Holy Spirit shall be fulfilled unto you, and ye shall go forth from Jerusalem unto all the tribes of Israel, and to the uttermost parts of the earth.

6. And having said these things, he lifted up his pure and holy hands and blessed them. And it came to pass that while he blessed them, he was parted from them, and a cloud, as the sun in brightness, received him out of their sight, and as he went up some held him by the feet and others worshipped him, falling to the earth on their faces.

7. And while they gazed steadfastly into heaven, behold two stood by them in white apparel, and said, Ye men of Israel, why stand ye gazing into thee, heaven; this same Jesus who is taken from you in a cloud, and as ye have seen him go into heaven, so shall he come again to the earth.

8. Then returned they unto Jerusalem from the Mount of Olives, which is from the city a Sabbath day's

journey. And as they returned they missed Mary Magdalene, and they looked for her, but found her not. And some of the disciples said, The Master hath taken her, and they marvelled and were in great awe.

9. Now it was midsummer when Jesus ascended into heaven, and he had not yet attained his fiftieth year, for it was needful that seven times seven years should be fulfilled in his life.

10. Yea, that he might be perfected by the suffering of all experiences, and be an example unto all, to children and parents, to the married and the celibates, to youth and those of full age, yea, and unto all ages and conditions of mortal life.

LECTION 95. 5. -There is no doubt that the "power" here referred to means the spiritual power which we read of as exercised by the followers of Jesus and other great prophets in all ages more or less. Taking the various accounts in the Gospel and ecclesiastical history as correct, miracles *(i.e.,* wondrous works wrought by the exercise of faith and will power and often by the uses of subtle forces of nature, quite natural, but seemingly supernatural to those in ignorance of these forces) were of frequent occurrences in those days, even as they are in these days, but better understood, false miracles being no proof of the non-existence of true ones. Often they would be the effect exercised on the minds and imaginations of vast numbers of the poor and aflicted, the diseased and suffering of humanity by faith in some great champions of the oppressed, themselves destroyed by the oppressor, yet realised by faith, if not by actual knowledge as still living and acting, with hands outstretched to heal and bless those who invoked their aid.

v. **9.** -From the testimony of the Jews, John viii. 57, A. *V.,* it appears that Jesus at that time was not far from fifty years of age, and this is supported by S. Irenmus, 120-200 A.D., who appeals to the gospel as received by those of his day and to all the elders as testifying the same," those who were conversant in Asia with John, the disciple of the Lord, affirming that John conveyed to them this tradition." "Some of them," he says again, "not only saw John but the other Apostles also, and heard the very same tradition from them. Bond bear testimony to the truth of the statement."

The Editor of this Gospel has been credibly informed by an esteemed friend of his, "a Syrian Bishop," and a relative of the late learned Pope Pius IX., that he frequently (in private) assured him that he firmly held this (as a private opinion), the present time (1870) not being yet ripe for a public declaration on this and similar subjects, now introduced into the notes to this and other publications of the O.A.

LECTION 95. 8. -Mary Magdalene was chosen by our Lord as a type of the Church, in her fallen condition, redeemed by His love, and would be fitly one of the first fruits taken to be with her Lord, as Ioseph and Mary were after. She was the constant companion of Iesus' Ministry, to him she ministered of her substance, she anointed him for his Ministry, and for his Burial. She was the last at the Cross, and the first at the Tomb, and to her aJone He gave the commission, " Go tell Peter," and wheresoever the Gospel was to be preached, her love and devotion to her Master were to be declared.

Lection 96

The Pouring Out Of The Spirit.

1. AND as the disciples were gathered together in the upper room when they returned from the Mount, they all continued with one accord in prayer and supplication, and their number was about one hundred and twenty.

2. And in that day James stood up and said; Men and brethren, it is known unto you how the Lord, before he left us, chose Peter to preside over us and watch over us in his Name; and how it must needs be that one of those who have been with us and a witness to his resurrection be chosen and appointed to take his place.

3. And they chose two called Barsabas and Matthias, and they prayed and said, Thou lord, who knowest the hearts of all men, shew which of these two thou hast chosen to take part in this Apostleship from which thou dost raise thy servant Peter to preside over us.

4. And they gave forth their lots, and the lot fell upon Matthias, and the Twelve received him, and he was numbered among the Apostles.

5. Then John and James separated Peter from their number by laying on of hands, that he might preside over them in the Name of the Lord, saying, Brother be thou as a hewn stone, sixsquared. Even thou, Petros, which art Petra, bearing witness to the Truth on every side.

6. And to the Apostles were given staves to guide their steps in the ways of truth, and crowns of glory withal; and to the Prophets burning lamps to shew light on the path and censers with fire; and to the Evangelists the book of the holy law to recall the people to the first principles; and to the Pastors were given the cup and platter to feed and nourish the flock.

7. But to none was given aught that was not given to all, for all were one priesthood under the Christ as their Master Great High Priest in the Temple of God; and to the Deacons were given baskets that they might carry therein the things needful for the holy worship. And the number was about one hundred and twenty, Peter presiding over them.

8. AND when the third day had fully come they were all with one accord in the one place, and as they prayed there came a sound from heaven as of a rushing mighty wind, and the room in which they were assembled was shaken, and it filled the place.

9. And there appeared cloven tongues of flame like fire, and sat upon the head of each of them. And they were all filled with the Holy Spirit and began to speak with tongues as the Spirit gave them utterance. And Peter stood up and preached the Law of Christ unto the multitude of all nations and tongues who were gathered together by the report of what had been seen and heard, each man hearing in his own tongue wherein he was born.

10. And of them that listened there were gathered unto the Church that day, three thousand souls, and they received the Holy Law, repented of their sins, and were baptized and continued stedfastly in the Apostles' fellowship and worship, and the Oblation and prayers.

11. And they who believed gave up their possessions, and had all things in common and abode together in one place, shewing the love and the goodness of God to their brothers and sisters and to all creatures, and working with their hands for the common weal.

12. And from these there were called twelve to be Prophets with the Apostles, and twelve to be Evangelists and twelve to be Pastors, and their Helps were added unto them, and Deacons of the Church Universal, and they numbered one hundred and twenty. And thus was the Tabernacle of David set up, with living men filled with goodness, even as the Master had shewn unto them.

13. And to the Church in Jerusalem was given James the Lord's brother for its president and Angel, and under him four and twenty priests in a fourfold ministry, and helpers and deacons also. And after six days many came together, and there were added six thousand men and women who received the holy Law of Love, and they received the word with gladness.

14. AND as they gathered together on the Lord's Day after the Sabbath was past, and were offering the holy Oblation, they missed Mary and Joseph, the parents of Jesus. And they made search but found them not.

15. And some of them said, Surely the Lord hath taken them away, as he did Magdalene. And they were filled with awe, and sung praises to God.

16. And the Spirit of God came upon the Apostles and the Prophets with them and, remembering what the Lord had taught them, with one voice they confessed and praised God, saying.

17. We believe in One God: the Infinite, the Secret Fount, the Eternal Parent: Of Whom are all things invisible and visible. The ALL in all, through all around all. The holy Twain, in whom all things consist; Who hath been, Who is, Who shall be.

18. We believe in one Lord our Lady, the perfect holy Christ: God of God, Light of light begotten. Our Lord, the Father, Spouse and Son. Our Lady, the Mother, Bride and Daughter. Three Modes in one Essence undivided: One Biune Trinity. That God may be manifest as the Father, Spouse and Son of every soul: and that every soul may be perfected as the Mother, Bride and Daughter of God

19. And this by ascent of the soul into the spirit and the descent of the spirit into the soul. Who cometh from heaven, and is incarnate of the Virgin ever blessed, in Jesu-Maria and every Christ of God: and is born and teacheth the way of life and suffereth under the world rulers, and is crucified, and is buried and descendeth into Hell. Who riseth again and ascendeth into glory; from thence giving light and life to all.

20. We believe in the Sevenfold Spirit of God, the Life-Giver: Who proceedeth from the holy Twain. Who cometh upon Jesu-Maria and all that are faithful to the light within: Who dwelleth in the Church, the Israel elect of God. Who cometh ever into the world and lighteth every soul that seeks. Who giveth the

Law which judgeth the living and the dead, Who speaketh by the Prophets of every age and clime.

21. We believe in One Holy Universal and Apostolic Church: the Witness to all truth, the Receiver and Giver of the same. Begotten of the Spirit and Fire of God: Nourished by the waters, seeds and fruits of earth. Who by the Spirit of Life, her twelve Books and Sacraments, her holy words and works: knitteth together the elect in one mystical communion and atoneth humanity with God. Making us partakers of the Divine Life and Substance: betokening the same in holy Symbols.

22. And we look for the coming of the Universal Christ: and the Kingdom of Heaven wherein dwelleth righteousness. And the holy City whose gates are Twelve: wherein are the Temple and Altar of God. Whence proceed three Orders in fourfold ministry: to teach all truth and offer the daily sacrifice of praise.

23. As in the inner so in the outer: as in the great so in the small. As above, so below: as in heaven so in earth. We believe in the Purification of the soul: through many births and experiences. The Resurrection from the dead: and the Life everlasting of the just. The Ages of Ages: and Rest in God for ever.—Amun.

24. And as the smoke of the incense arose, there was heard the sound as of many bells, and a multitude of the heavenly host praising God and saying:

25. Glory, honour, praise and worship be to God; the Father,, Spouse, and Son: One with the Mother, Bride and Maid: From Whom proceedeth the Eternal Spirit: By whom are all created things. From the Ages of Ages. Now: and to the Ages of Ages—Amun—Alleluia, Alleluia, Alleluia.

26. And if any man take from, or add, to the words of this Gospel, or hide, as under a bushel, the light thereof, which is given by the Spirit through us, the twelve witnesses chosen of God, for the enlightenment of the world unto salvation: Let him be Anathema Maranatha, until the coming of Christ Jesu-Maria, our Saviour, with all the Holy Saints.

27. For them that believe, these things are true. For them that believe not, they are as an idle tale. But to those with perceiving minds and hearts, regarding the spirit rather than the letter which killeth, they are spiritual verities.

28. For the things that are written are true, not because they are written, but rather they are written because they are true, and these are written that ye may believe with your hearts, and proclaim with your mouths to the salvation of many. *Amen.*

Here endeth the Holy Gospel of the Perfect Life of Jesu-Maria, the Christ, the Son of David after the Flesh, the Son of God after the Spirit. Glory be to God by Whose power and help it has been written.

LECTION 96. 1. -This number, 120, has many mystic significances, and was foreshadowed by the number of souls saved in the Ark at the Flood ("The Original Genesis"), which included 48 *(i.e.,* double 7 + 34) + 72, a number of deep mystic significance.

v. **2.** -The manifestations described here have been repeated in modern times. What God does in one age, whether by angels, spirits, or adepts in the flesh, the same unchanging God repeats in another. Whether the miracle respecting the preaching of Peter took place in the persons of the Apostles, or in their hearers, we have no means of ascertaining, but the fact remains. Most probably in the hearing of the hearers, so that each was enabled spiritually to understand. or else all were moved to speak and to hear in a tongue common to all. **Cont**

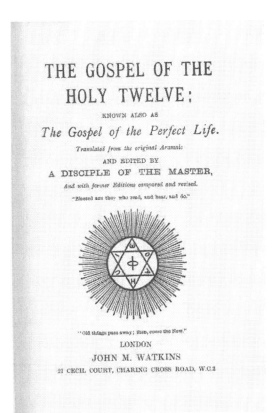
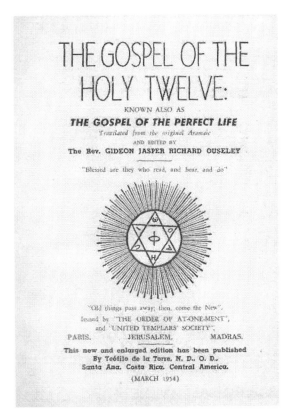

1956

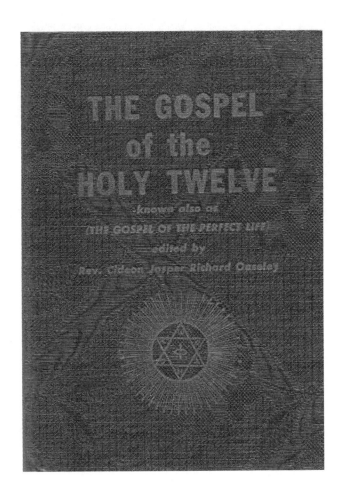

The Gospel Of The Holy 12

Sample Pages from Jain's Early Publication

Original Front Cover by Jain

THE GOSPEL OF THE HOLY TWELVE

This rare Essene Bible was found in an urn in Tibet in a Buddhist monastery where it had been laying for 1800 years until it was recently rediscovered and translated from the ancient Aramaic (the language Jesus spoke) into English. Of the 4 translators, one of them was the famed Emmanuel Swedenborg (died 1772), the Swedish mystic and seer.

My particular interest in this manuscript is for 5 reasons:

1: I first discovered this book at the Adyar bookshop, Sydney, around the time of 1977 when I was 20 years of age and it was protruding from a top shelf. As I jumped to push the book back in place, I suddenly decided to grab it. I normally would not be standing in front of Christian material. This book changed my life.

2: As I flipped through the pages, this book had many images of Magic Squares of 3x3, 7x7 and 11x11 and sacred symbols like Pythagoras' Triangle, Egyptian Ankhs, Star of Davids, Alpha and Omega symbols, Ground Plan of the Christian Church, Labyrinth etc.

3: The Lord's Prayer, in Lection XIX, in the chapter: "Iesus Teaches Concerning Prayer" was the first book I had encountered that honored both the Father and the Mother: It begins:
"OUR FATHER-MOTHER Who Art Above And Within: Hallowed Be Thy Sacred Name In Twofold Trinity . In Wisdom, Love and Equity Thy Kingdom Come To All...". I believe that the words "In Twofold Trinity" is a clear reference to the Phi Ratio which examines a Trinity of 3 parts into a specific twofold partitioning of 2 parts, one is the larger segment and the other is the smaller segment).

4: There is the most beautiful story about a man born blind from birth who Jesus heals. It is called: "The Examination of Him Who Was Born Blind" (Lection LIV). This man becomes devoted to Jesus' teachings and asks many questions. It continues to describe the Divine Kingdom in terms of the Magic Square of 7x7 and is referred to as THE PARABLE OF THE SEVEN PALMS. From this material I have been inspired to create what I call THE THEATRE OF THE HOLY NUMBERS which is a creative re-enactment of this biblical scene. The Magic Square of 11x11, like the 7x7 is important to the Essenes as the central crosses in both also add up to the Magic Sum or Constant of the Rows and Columns. In the Brief Commentary, the Magic Square of Eleven Perfected is subtitled: "Ecclesiae Militantis Sigilum". which means the Magic Square of 11x11 is the Churches' Military Seal of Power or Defence.

5: This is a very rare book with no ISBN. It appears to be from the same body of works that documented the missing 18 years of Jesus' life deleted from the modern bibles, and is essential reading.

Original Back Cover by Jain

1) - ABOUT THE BOOK: "THE GOSPEL OF THE HOLY 12"

Contains 196 photocopied pages, colour front and back cover, $50, wire bound, read and bound in the landscape format. The book has not been computerized, I have kept the original pages as they appeared in a 1974 reprinted version. Perhaps a later version will have it all computerized. Very fine print but still readable.

The front cover only says:
"Translated from the original Aramaic and edited by a Disciple of the Master..."

This is a very rare book with no ISBN; and I have not seen another in print, nor met anyone that has heard of it. It appears to be from the same body of works that documented the missing 18 years of Jesus' life deleted from the modern bibles, and is essential reading.

---

2) - BACK COVER BLURB

This rare Essene Bible was found in an urn in Tibet in a Buddhist monastery where it had been laying for 1800 years until it was recently rediscovered and translated from the ancient Aramaic (the language Jesus spoke) into English. Of the 4 translators, one of them was the famed Emmanuel Swedenborg (died 1772), the Swedish mystic and seer.

My particular interest in this manuscript is for 4 reasons:

1: I first discovered this book at the Adyar bookshop, Sydney, around the time of 1977 when I was 20 years of age and it was protruding from a top shelf. As I jumped to push the book back in place, I suddenly decided to grab it. I normally would not be standing in front of Christian material. This book changed my life.

2: As I flipped through the pages, this book had many images of Magic Squares of 3x3, 7x7 and 11x11 and sacred symbols like Pythagoras' Triangle, Egyptian Ankhs, Star of Davids, Alpha and Omega symbols, Ground Plan of the Christian Church, Labyrinth etc.

3: The Lord's Prayer, in Lection XIX, in the chapter: "Jesus Teaches Concerning Prayer" was the first book to honour both the Father and the Mother: It begins:

"OUR FATHER-MOTHER Who Art Above And Within: Hallowed Be Thy Sacred Name In Twofold Trinity. In Wisdom, Love and Equity Thy

Kingdom Come To All...".

I believe that the words "In Twofold Trinity" is a clear reference of the Phi Ratio which examines a Trinity of 3 parts into a specific twofold partitioning of 2 parts, one is the larger segment and the other is the smaller segment.

4: There is the most beautiful story about a man born blind from birth who Jesus heals. It is called: "The Examination of Him Who Was Born Blind" (Lection LIV). This man becomes devoted to Iesus' teachings and asks many questions. It continues to describe the Divine Kingdom in terms of the Magic Square of 7x7 and is referred to as THE PARABLE OF THE SEVEN PALMS. (The Yantram on the front cover is a Magic Square of 7x7 rotated 8 times, inspired from this Parable). From this material I have been inspired to create what I call THE THEATRE OF THE HOLY NUMBERS which is a creative re-enactment of this biblical scene.

The Magic Square of 11x11, like the 7x7 is important to the Essenes as the central crosses in both also add up to the Magic Sum or Constant of the Rows and Columns. In the Brief Commentary, the Magic Square of Eleven Perfected is subtitled: "Ecclesiae Militantis Sigilum". which means the magic Square of 11x11 is the Churches' Military Seal of Power or Defense.

A TRUE LIKENESS OF

OUR SAVIOUR.

THE GOSPEL OF THE HOLY TWELVE;

KNOWN ALSO AS

The Gospel of the Perfect Life.

Translated from the original Aramaic

AND EDITED BY

A DISCIPLE OF THE MASTER,

And with former Editions compared and revised.

"Blessed are they who read, and hear, and do."

Reprinted 1974
By

HEALTH RESEARCH
P.O. BOX 70
MOKELUMNE HILL, CALIFORNIA 95245

Original Front Cover

Notice the cryptic Magic Square of 3x3 where all the sums of rows and columns and diagonals = 96

xii.   TO THE READER.

As an aid to a higher Christianity these fragments of a fuller Gospel are now presented, giving us the Feminine tenderness as the Masculine strength of the Perfect Christ.

The greater and more important portion of these reminiscences have formed the groundwork and basis of various teachings issued by the Order of At-one-ment since 1881, when it was incepted, and are now for the first time given in their entirety, throwing additional light on the real doctrine of Jesus, or elucidating of the contents of the canonical Gospels as commonly received, retaining the translation of the A.V. wherever possible, or sufficiently clear. It will be for the Church of the Future when revising the entire Scriptures to give it its primary place, the original and complete "Gospel of the Holy Christ," using the others for a confirmation from four other witnesses that every word may be established to them who are not in a condition to receive the goodness, purity and truth of the former.

Like all other inspired writings (but not necessarily infallible in every word) these writings from within the Veil must be taken on their own internal evidence of a Higher Teaching. For inspiration of the Spirit no more necessarily implies infallibility than the divine breath of life inbreathed by man, necessarily implies freedom from all accidents, diseases or miseries incidental to mortal life.

It is a faithless and perverse generation, as of old, that seeks for signs, and to them saith the Spirit, "there shall no sign be given," for were the very writers of this Gospel raised from the dead, and were they to testify to their authorship, they would not believe, unbelieving critics would still ask for a sign, and the more they were given the more they would ask in the hardness of their hearts. The sign is The Truth—the pure in heart they shall see it.

(xiii.)

31	36	29
30	32	34
35	28	33

XCVI.

CONTENTS.

		PAGE
	PROLOGUE.	
LECTION		
I.	Parentage and Conception of Iohn the Baptist ..	1
II.	Immaculate Conception of Iesus the Christ ..	2
III.	Nativity of Iohn the Baptist ..	5
IV.	Nativity of Iesus the Christ ..	6
V.	Manifestation of Iesus to the Magi ..	8
VI.	Childhood and Youth of Iesus the Christ. He delivereth a lion from the hunters	11
VII.	The Preaching of Iohn the Baptist ..	14
VIII.	The Baptism of Iesus-Maria, the Christ..	15
IX.	The Four Temptations ..	17
X.	Joseph and Mary make a feast unto Iesus. Andrew and Peter find Iesus ..	18
XI.	The Anointing by Mary Magdalene ..	20
XII.	The Marriage in Cana. The Healing of the Nobleman's Son ..	21
XIII.	The First Sermon in the Synagogue of Nazareth ..	23
XIV.	The Calling of Andrew and Peter. The Teaching of Cruelty in Animals. The Two Rich Men ..	24
XV.	The Healing of the Leper and the Palsied. The Deaf man who denied that others could Hear ..	25
XVI.	Calling of Matthew. Parable of the New Wine in Old Bottles ..	27
XVII.	Iesus Sendeth Forth the Twelve and their Fellows ..	28

CONTENTS.

LECTION		PAGE
XVIII.	Iesus Sendeth Forth the Two and Seventy	30
XIX.	Iesus Teacheth how to Pray. Error even in Prophets	31
XX.	The Return of the Two and Seventy	32
XXI.	Iesus rebuketh Cruelty to a Horse. Condemneth the Service of Mammon. Blesseth Infants	33
XXII.	Restoration of Iairus' Daughter	35
XXIII.	Iesus and the Samaritan Woman	36
XXIV.	Iesus Denounces Cruelty. Healeth the Sick	37
XXV.	The Sermon on the Mount (Part I.)	39
XXVI.	Sermon on the Mount (Part II.)	41
XXVII.	Sermon on the Mount (Part III.)	43
XXVIII.	Iesus releaseth the Rabbits and Pigeons	45
XXIX.	He feedeth Five Thousand with Six Loaves, and Seven Clusters of Grapes. Heals the Sick. Jesus Walketh on the Water	47
XXX.	The Bread of Life and the Living Vine	49
XXXI.	The Bread of Life and the Living Vine. Iesus Teacheth the Thoughtless Driver	50
XXXII.	God the Food and Drink of All	52
XXXIII.	By the Shedding of Blood of Others is no Remission of Sins	53
XXXIV.	Love of Iesus for All Creatures. His Care for a Cat	55
XXXV.	The Good Law. The Good Samaritan. Mary and Martha. On Divine Wisdom	56
XXXVI.	The Woman taken in Adultery. The Pharisee and the Publican	58
XXXVII.	The Regeneration of the Soul	59
XXXVIII.	Iesus Condemneth the Ill-Treatment of Animals	60
XXXIX.	The Kingdom of Heaven (Seven Parables)	61
XL.	Iesus Expoundeth the Inner Teaching to the Twelve	63
XLI.	Iesus setteth free the Caged Birds. The Blind Man Who denied that others saw	64
XLII.	Iesus teacheth Concerning Marriage. The Blessing of Children	66
XLIII.	Iesus Teacheth Concerning the Riches of this World and the Washing of Hands and Unclean Meats	67
XLIV.	The Confession of the Twelve. Christ the True Rock	69
XLV.	Seeking for Signs. The Unclean Spirit	71
XLVI.	The Transfiguration on the Mount, and the Giving of the Law	72
XLVII.	The Spirit giveth Life. The Rich Man and the Beggar	75
XLVIII.	Iesus feedeth One Thousand with Five Melons. Healeth the Withered Hand on the Sabbath Day. Rebuketh Hypocrisy	77
XLIX.	The True Temple of God	78
L.	Christ the Light of the World	80
LI.	The Truth Maketh Free	81
LII.	The Pre-existence of Christ	83
LIII.	Iesus Healeth the blind on the Sabbath. Iesus at the Pool of Siloam	84
LIV.	The Examination of the Blind Man. A Living Type of the House of God	85
LV.	Christ the Good Shepherd	88
LVI.	The Raising of Lazarus from his Sleep in the Tomb	90
LVII.	Concerning little Children. Forgiveness of others. Parable of the Fishes	92
LVIII.	Divine Love to the Repentant	94
LIX.	Iesus Forewarneth His Disciples. Glad Tidings to Zacchæus	97
LX.	Iesus Rebuketh Hypocrisy	98
LXI.	Iesus Foretelleth the End	101
LXII.	Parable of the Ten Virgins	103
LXIII.	Parable of the Talents	104
LXIV.	Iesus Teacheth in the Palm Circle. The Divine Life and Substance	105
LXV.	The Last Anointing by Mary Magdalene. Neglect not Present time	107
LXVI.	Iesus again Teacheth his Disciples concerning the Nature of God. The Kingdom of God. The Two in One	108

contents continued:

CONTENTS.

LECTION		PAGE
LXVII.	The Last Entry into Jerusalem. The Sheep and the Goats..	109
LXVIII.	The Householder and the Husbandmen. Order out of Disorder	111
LXIX.	The Christ within the Soul. The Resurrection and the Life. Salome's Question	113
LXX.	Iesus Rebuketh Peter's Haste	115
LXXI.	The Cleansing of the Temple	117
LXXII.	The Many Mansions in the One House	118
LXXIII.	Christ the True Vine	120
LXXIV.	Iesus foretelleth Persecutions	121
LXXV.	The Last Paschal Supper..	123
LXXVI.	Washing of Feet. The Eucharistic Oblation	125
LXXVII.	The Agony in Gethsemane	129
LXXVIII.	The Betrayal	131
LXXIX.	The Hebrew Trial before Caiaphas	132
LXXX.	The Sorrow and Penance of Iudas	134
LXXXI.	The Roman Trial before Pilate ..	135
LXXXII.	The Crucifixion	138
LXXXIII.	The Burial of Iesus	141
LXXXIV.	The Resurrection of Iesus	142
LXXXV.	Iesus appeareth to Two at Emmaus	144
LXXXVI.	Iesus appeareth in the Temple. Blood Sacrifices Cease	145
LXXXVII.	Iesus appeareth to the Twelve	146
LXXXVIII.	The Eighth Day after the Resurrection	148
LXXXIX.	Iesus appeareth at the Sea of Galilee	149
XC.	What is Truth?	151
XCI.	The Order of the Kingdom (Part I.)	152
XCII.	The Order of the Kingdom (Part II.)	154
XCIII.	The Order of the Kingdom (Part III.)	156
XCIV.	The Order of the Kingdom (Part IV.)	157
XCV.	The Ascension of Christ	159
XCVI.	The Pouring out of the Spirit. The taking of Mary and Joseph..	161
The Epistle of Apollos the Prophet		165
Foreword		168
A Brief Commentary		169

IN THE NAME OF THE ALL HOLY. AMUN.

HERE beginneth the Gospel of the perfect Life of Iesu-Maria, the Christ, the offspring of David through Ioseph and Mary after the flesh, and the Son of God, through Divine Love and Wisdom, after the Spirit.

PROLOGUE.

From the Ages of ages is the eternal Thought, and the Thought is the Word, and the Word is the Act, and these Three are one in the Eternal Law, and the Law is with God and the Law proceeds from God. All things are created by Law and without It is not anything created that existeth. In the Word is Life and Substance, the Fire and the Light. The Love and the Wisdom, are One for the Salvation of all. And the Light shineth in darkness and the darkness concealeth it not. The Word is the one Life-giving Fire, which shining into the world becometh the fire and light of every soul that entereth into the world. I am in the world, and the world is in Me, and the world knoweth it not. I come to my own House, and my friends receive Me not. But as many as receive and obey, to them is given the power to become the sons and daughters of God, even to them who believe in the Holy Name, who are born—not of the will of the blood and flesh, but of God. And the Word is incarnate and dwelleth among us, whose Glory we beheld, full of Grace. Behold the Goodness, and the Truth and the Beauty of God!

LECTION XIII.

His First Sermon in the Synagogue.

1 AND Iesus came to Nazareth, where he had been brought up: and, as his custom was, he went into the synagogue on the sabbath day, and stood up for to read. And there was delivered unto him the roll of the prophet Esaias.

2 And when he had opened the roll, he found the place where it was written. The Spirit of the Lord is upon me, because he hath anointed me to preach the gospel to the poor; he hath sent me to heal the brokenhearted, to preach deliverance to the captives and recovering of sight to the blind, to set at liberty them that are bound. To preach the acceptable year of the Lord.

3 And he closed the roll, and gave it again to the minister, and sat down. And the eyes of all them that were in the synagogue were fastened on him. And he began saying unto them, This day is this scripture fulfilled in your ears. And all bare him witness, and wondered at the gracious words which proceeded out of his mouth. And they said, Is not this Ioseph's son?

4 And some brought unto him a blind man to test his power, and said, Rabbi, here is a son of Abram blind from birth. Heal him as thou hast healed Gentiles in Egypt. And he, looking upon him, perceived his unbelief and the unbelief of those that brought him, and their desire to ensnare him. And he could do no mighty work in that place because of their unbelief.

5 And they said unto him, Whatsoever we have heard done in Egypt, do also here in thy own country. And he said, Verily I say unto you, No prophet is accepted in his own home or in his own country, neither doth a physician work cures upon them that know him.

6 And I tell you of a truth, many widows were in Israel in the days of Elias, when the heaven was shut up three years and six months, when great famine was throughout all the land. But unto none of them was Elias sent, save unto Sarepta, a city of Sidon, unto a woman that was a widow.

7 And many lepers were in Israel in the time of Eliseus the prophet; and none of them was cleansed, saving Naaman the Syrian.

8 And all they in the synagogue, when they heard these things, were filled with wrath. And rose up, and thrust him out of the city, and led him unto the brow of

The Original Lord's Prayer from this Essene Text:

2 pages:

LECTION XVIII.

THE SENDING OF THE TWO AND SEVENTY.

1 AFTER these things the Lord appointed two and seventy also, and sent them two and two before his face into every city and place of the tribes whither he himself would come.

2 Therefore said he unto them, The harvest truly is great, but the labourers are few, pray ye therefore the Lord of the harvest that he would send forth labourers into the harvest.

3 Go your ways, behold I send you forth as lambs among wolves. Carry neither purse, nor scrip, nor shoes, and salute no man by the way.

4 And into whatsoever house ye enter, first say, Peace be to this house. And if the spirit of peace be there your peace shall rest upon it, if not it shall turn to you again.

5 And into whatsoever city ye enter, and they receive you, eat such things as are set before you without taking of life. And heal the sick that are therein, and say unto them, The kingdom of God is come nigh unto you.

6 And in the same house remain, eating and drinking such things as they give without shedding of blood, for the labourer is worthy of his hire. Go not from house to house.

7 But into whatsoever city ye enter and they receive you not, go your ways out into the streets of the same and say, Even the very dust of your city, which cleaveth on us, we do wipe off against you, notwithstanding be ye sure of this, that the kingdom of God is come nigh unto you.

8 Woe unto thee, Chorazin! woe unto thee, Bethsaida! for if the mighty works had been done in Tyre and Sidon, which have been done in you, they had a great while ago repented, sitting in sackcloth and ashes. But it shall be more tolerable for them in the judgment than for you.

9 And thou, Capernaum, which art exalted to heaven shalt be thrust down to hades. They that hear you, hear also me; and they that despise you, despise also me; and they that despise me, despise Him that sent me. But let all be persuaded in their own minds.

10 AND again Iesus said unto them: Be merciful, so shall ye obtain mercy. Forgive others, so shall ye be forgiven. With what measure ye mete, with the same shall it be meted unto you again.

11 As ye do unto others, so shall it be done you. As ye give, so shall it be given unto you. As ye judge others, so shall ye be judged. As ye serve others, so shall ye be served.

12 For God is just, and rewardeth every one according to their works. That which they sow they shall also reap.

LECTION XIX.

IESUS TEACHES CONCERNING PRAYER.

1 As Iesus was praying in a certain place on a mountain, some of his disciples came unto him, and one of them said, Lord teach us how to pray. And Iesus said unto them, When thou prayest enter into thy secret chamber, and when thou hast closed the door, pray to Abba Amma Who is above and within thee, and thy Father-Mother Who seest all that is secret shall answer thee openly.

2 But when ye are gathered together, and pray in common, use not vain repetitions, for your heavenly Parent knoweth what things ye have need of before ye ask them. After this manner therefore pray ye:—

3 Our Father-Mother Who art above and within: Hallowed be Thy sacred Name in twofold Trinity. In Wisdom, Love and Equity Thy Kingdom come to all. Thy will be done, As in Heaven so in Earth. Give us day by day to partake of Thy holy Bread, and the fruit of the living Vine. As Thou dost forgive us our trespasses, so may we forgive others who trespass against us. Shew upon us Thy goodness, that to others we may shew the same. In the hour of temptation, deliver us from evil.

4 For Thine are the Kingdom, the Power and the Glory: From the Ages of ages, Now, and to the Ages of ages. Amun.

5 And wheresoever there are seven gathered together in My Name there am I in the midst of them; yea, if only there be three or two; and where there is but one who prayeth in secret, I am with that one.

6 Raise the Stone, and there thou shalt find me. Cleave the wood, and there am I. For in the fire and in the water even as in every living form, God is manifest as it's Life and it's Substance.

32 THE GOSPEL OF THE HOLY TWELVE;

7 And the Lord said, If thy brother hath sinned in word seven times a day, and seven times a day hath made amendment, receive him. Simon said to him, Seven times a day?

8 The Lord answered and said to him, I tell thee also unto seventy times seven, for even in the Prophets, after they were anointed by the Holy Spirit, utterance of sin was found.

9 Be ye therefore considerate, be tender, be ye pitiful, be ye kind, not to your own kind alone, but to every creature which is within your care, for ye are to them as gods, to whom they look in their need. Be ye slow to anger for many sin in anger which they repented of, when their anger was past.

10 And there was a man whose hand was withered and he came to Iesus and said, Lord, I was a mason seeking sustenance by my hands, I beseech thee restore to me my health that I may not beg for food with shame. And Iesus healed him, saying, There is a house made without hands, seek that thou mayest dwell therein.

LECTION XX.

The Return of the Two and Seventy.

1 And after a season the two and seventy returned again with joy, saying, Lord, even the demons are subject unto us through thy name.

2 And he said unto them, I beheld Satan as lightning fall from heaven.

3 Behold I give unto you power to tread on serpents and scorpions, and over all the power of the enemy; and nothing shall by any means hurt you. Notwithstanding in this, rejoice not, that the spirits are subject unto you; but rather rejoice, because your names are written in heaven.

4 In that hour Iesus rejoiced in spirit, and said, I thank thee, Holy Parent of heaven and earth, that thou hast hid these things from the wise and prudent, and hast revealed them unto babes: even so, All Holy, for so it seemed good in thy sight.

5 All things are delivered to me of the All Parent: and no man knoweth the Son who is the Daughter, but the All Parent; nor who the All Parent is, but the Son even

OR, THE GOSPEL OF THE PERFECT LIFE. 33

the Daughter, and they to whom the Son and the Daughter will reveal it.

6 And he turned him unto his disciples, and said privately, Blessed are the eyes which see the things that ye see. For I tell you, that many prophets and kings have desired to see those things which ye see, and have not seen them; and to hear those things which ye hear, and have not heard them.

7 Blessed are ye of the inner circle who hear my word and to whom the mysteries are revealed, who give to no innocent creature the pain of prison or of death, but seek the good of all, for to such is everlasting life.

8 Blessed are ye who abstain from all things gotten by bloodshed and death, and fulfil all righteousness: Blessed are ye, for ye shall attain to Beatitude.

LECTION XXI.

Iesus Rebuketh Cruelty to a Horse.

1 And it came to pass that the Lord departed from the City and went over the mountains with his disciples. And they came to a mountain whose ways were steep and there they found a man with a beast of burden.

2 But the horse had fallen down, for it was overladen, and he struck it till the blood flowed. And Iesus went to him and said: "Son of cruelty, why strikest thou thy beast? Seest thou not that it is too weak for its burden, and knowest thou not that it suffereth?"

3 But the man answered and said: "What hast thou to do therewith? I may strike it as much as it pleaseth me, for it is mine own, and I bought it with a goodly sum of money. Ask them who are with thee, for they are of mine acquaintance and know thereof."

4 And some of the disciples answered and said: Yea, Lord, it is as he saith, We have seen when he bought it. And the Lord said again: "See ye not then how it bleedeth, and hear ye not also how it waileth and lamenteth?" But they answered and said: "Nay, Lord, we hear not that it waileth and lamenteth?"

5 And the Lord was sorrowful, and said: "Woe unto you because of the dulness of your hearts, ye hear not how it lamenteth and crieth unto the heavenly Creator for mercy, but thrice woe unto him, against whom it crieth and waileth in its pain."

---

continued: Jain's calligraphy applied to this Lord's prayer:

The Book Of Magic Squares, by Jain, Volume 1

"Out of the darkness ariseth the light."

"There are many things which I have to say unto you, but ye cannot bear them now. Howbeit the Spirit of Truth whom I will send unto you from above, shall guide you into all truth, and bring all things to your remembrance whatsoever I have said unto you."

"After my departure there will arise the ignorant and the crafty, and many things will they ascribe unto Me that I never spake, and many things which I did speak will they withhold, but the day will come when the clouds shall be rolled away, and the Sun of Righteousness shall shine forth with healing in his wings."

"I am the Way, the Truth, the Life. The doctrine which I teach is that which I am. I am It, and It is Me."

"There are also many things which Iesus did, the which if they should be written, I suppose that the world itself could not contain the books that should be written." — S. Iohn (A.V.).

"In dreams and visions of the night when deep sleep falleth upon men, then doth the All Wise open their ears and send to them instruction." — Book of Job.

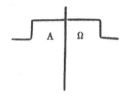

"Unitas per Libram crucis."

EXPLANATORY PREFACE
TO THE NEW AND COMPLETE EDITION
(REVISED AND ENLARGED.)

This "Gospel of the Holy Twelve" (Evangelists) of the Christian Dispensation is one of the most ancient and complete of early Christian fragments, preserved in one of the Monasteries of the Buddhist monks in Thibet, where it was hidden by some of the Essene community for safety from the hands of corrupters and now for the first time translated from the Aramaic. The contents clearly show it to be an early Essenian writing. This ancient community of the Jewish Church called Yessenes, Iessenes, Nazarites, or Nazirs, strongly resembling the Therapeutæ, and the Buddhists, who practised community of goods, daily ablutions, daily worship, and renounced flesh eating, and strong drink and the sacrifice of animals, and the doctrine of "atonement" for the sins of some by the vicarious and involuntary suffering of others, as held by the Pharisees and Sadducees, and by the heathen before them; thus preparing the way for those Orders and Communities of men and women which have since arisen throughout the East and West, like cities set on a hill, to shew the more perfect way to Christians living in the world, notably those of S. Basil in the East, and S. Benedict in the West, and, with them, the Carthusians and the Franciscans, and before them all, the Carmelites (who had their headquarters on Mount Carmel) to whom they are similar in their customs, and even their dress, if not altogether identical with them, tracing their origin to Elias, abstaining from all flesh meats and strong drinks, whose symbol was, it is said, an iron cross in a circle, and among the animals, the Lamb and the Dove their special emblems. See Philo (*in Loco*) or Kitto's Cyclopædia (art, Essenes), also Arthur Lillie's "Christianity and Buddhism."

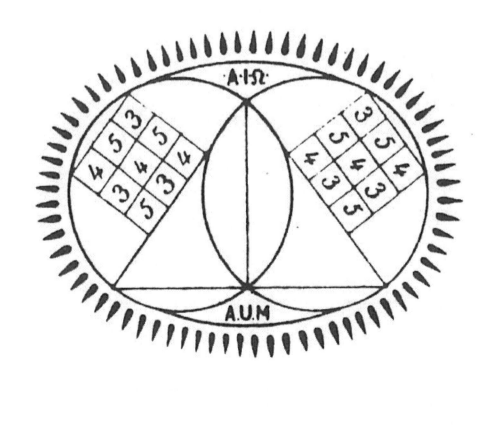

EXPLANATORY PREFACE.

For long ages, since the destruction of Jerusalem, the aspect of God, as the Eternal Mother ("Holy Spirit"), One with the Eternal Father, has been concealed from our eyes by clouds of human tradition, error and superstition, but now the Sun of Righteousness (the Divine Mother), shines in its fulness from behind the clouds of darkness that have so long hidden it from view, and it shines alike on believer and unbeliever, on those who see and feel its warmth, and on those who will to remain in darkness, and perish for want of light. It is written, "Behold I come from behind the clouds and every eye shall see Me whom they have pierced, and all of the earth who see, shall wail because of their iniquities." The world is deaf to the beloved Voice, and hears not, blind and sees not, but only the things which make for its own lusts. As of yore, approaching to the Holy City, Jesus wept, so now to modern Christendom He seems to say, "Ierusalem, Ierusalem, thou that stonest the prophets and revilest them that are sent unto thee, how often would I have gathered thee as a hen gathereth her chicken under her wings, but ye would not! Oh that thou hadst listened in this thy day to the things which make (through righteousness) for thy peace, but ye would not! Behold the day cometh, when thine enemies shall cast a trench before thee and surround thee on every side, and burn thee with fire, leaving not one stone upon another. Behold now is your house left unto you desolate, and ye shall not see Me henceforth, till ye shall say, Blessed are They who come in the Name of the Holy One."

THE EDITORS OF THE GOSPEL OF THE HOLY TWELVE.

TO THE READER.

THE all-pitying love of our Saviour embraces not only mankind, but also the "lower" creatures of God, sharers with us of the one breath of life, and with us on the one road of ascent to that which is higher. Never has the Providence with which the All-Merciful watches over the animals "unendowed with reason," as well as over "reason-endowed" man, been more impressively brought home to us than in the saying of Iesus: "Are not five sparrows sold for two farthings, and not one of them is forgotten before God?"

How were it possible to think otherwise, than that the Saviour would have had pity and compassion on the creatures who must bear their pain in silence. Would it not seem to us like a blasphemy if we were to hear it said that Iesus or Mary would have beheld, without pity or succour the illtreatment of helpless animals? Nay, certainly, when our Saviour brought redemption to a world sunk in selfishness, hard-heartedness and misery, and proclaimed the Gospel of an all-embracing love, there was, surely, a share in this redemption for all suffering creatures, since when man opened his heart to this Divine Love, there was no room left for pitiless hardness towards the other creatures of God, who have, like man, been called into life with the capacity of enjoyment and of suffering. They bear the marks of the Redeemer, who practises this all-pitying love. And how little it is that the minimum of Christian compassion for helpless creatures demands of us! Only to inflict on them no torture, and to help them when in trouble, or appeal to us for succour, and when, of necessity, we take their life, to let it be a speedy death with the least pain; a gentle sleep. But, alas! how little are we penetrated with these divine lessons of mercy and compassion. How many grievous tortures are inflicted on helpless creatures under the pretence of science, or to gratify an unnatural appetite, or cruel lusts, or the promptings of vanity.

THE GOSPEL OF THE HOLY TWELVE.

LECTION I.

The Parentage and Conception of John the Baptist.

1 There was in the days of Herod, the king of Judea, a certain priest named Zacharias, of the course of Abia; and his wife was of the daughters of Aaron, and her name was Elisabeth.

2 And they were both righteous before God, walking in all the commandments and ordinances of the Lord blameless. And they had no child, because that Elisabeth was barren, and they both were now well stricken in years.

3 And it came to pass, that while he executed the priest's office before God in the order of his course, according to the custom of the priest's office, his lot was to burn incense when he went into the temple of IOVA. And the whole multitude of the people were praying without at the time of the offering of incense.

4 And there appeared unto him an angel of the Lord standing over the altar of incense. And when Zacharias saw, he was troubled, and fear fell upon him. But the angel said unto him, Fear not, Zacharias, for thy prayer is heard; and thy wife Elisabeth shall bear thee a son, and thou shalt call his name John.

5 And thou shalt have joy and gladness; and many shall rejoice at his birth; for he shall be great in the sight of the Lord, and shall neither eat flesh meats, nor drink strong drink; and he shall be filled with the Holy Spirit, even from his mother's womb.

6 And many of the children of Israel shall he turn to the Lord their God; And he shall go before him in the

A sample page with the Crown of Thorns, that says:

"The sorrows of the earth encompass my heart, yea,
they pierce it with thorns
but thou wilt yet fill the earth with love
and crown it with gladness".

A Brief Commentary:

LECTION LIV. 14-16.—The meaning of these words and this action is very obscure, but if we describe the magic square of 7, it seems to make it intelligible as the mystic symbol of him who regarded everything by number and by measure, and which seems to have reference to the period of his mortal life, 49 years, as well as the number of the Council, Cardinals and Priests of the Church universal, 48, presided over by its Head, 49, which the action of Iesus seemed to symbolize, and in a way, foreshadow.

LECTION LIV. 17-20.—Here we have the original words of Christ, from which Paul adopted his simile in Rom. xii., and in 1 Cor. xii.

MAGIC SQUARE OF SEVEN.
SIGILLUM XTI.

22	47	16	41	10	35	4
5	23	48	17	42	11	29
30	6	24	49	18	36	12
13	31	7	25	43	19	37
38	14	32	1	26	44	20
21	39	8	33	2	27	45
46	15	40	9	34	3	28

21 Summations of 175.   Sum total of terms, 225.

This most ancient seal of XT—the High Priest, the At-one-er and At-one-ment of all things, the Lord our Lady Iesu-Maria—is one of the most ancient and sacred symbols of the Xn religion, extending back to pre-Xn antiquity, and symbolising the restoration of the primeval order to the universe, bringing order out of confusion. In this most curious and mystical combination of

The Magic Square of 7x7 describing "The Parable of the 7 Palms". The unusual part of this Magic Square is that the central crosses of 7 combined cells also add up to the same as the rows, columns and diagonals, which is not seen in any other magic squares. It therefore contained a mystical significance.

help me, whistled to it, whereupon it turned round and looked at us as though to see what we wanted. Then *I put away my gun, for I had not the heart to shoot.* It would have been like killing a house-dog." (See also "Darwin's Voyage on H.M.S. *Beagle*," where a similar testimony is given). If man deprives his essential knowlege of good, he disarranges in himself all that part of creation that is within him, and becomes lower than the beasts of prey.

LECTION LII. 4.—The testimony of those who saw and knew Iesus as to his age, has been strangely ignored by writers of Biblical history and by the Church in general. This matter is briefly discussed elsewhere in these Notes, and deserves the attention of every student and thoughtful person. (See Notes liv. 14-16; xcv. 9.)

LECTION LIII. 3.—The healing of the blind by means of clay mingled with saliva is mentioned by ancient physicians. Vespasian is said to have cured by this means. This shows that Iesus did not hesitate to employ natural remedies, when they were likely to effect their purpose.

LECTION LIV. 1-13.—The wrangling of the Pharisees over this case of healing has its parallels in our times in the Churches which assign to the devil all that they cannot comprehend, and cast out the Healer as a sinner and a heretic, denying the power of God in Man.

LECTION LIV. 14.—This is one of those "parables and dark sayings" of him who spake as never man spake. The words taken literally suggest to the mind a perfect crystal sphere, and by correspondence, a perfect man or woman—in modern phrase "an all rounder," one who views things not from one side only, but from every side. There are many who keep the law in one or more points, but neglect all the rest; or keep it in all points but the one which is against their own particular failing—who "compound for sins they are inclined to, by damming those they have no mind to." But few are they who teach, and still fewer who practice an all round obedience to the laws of Christ. Many are they who loudly condemn one or more forms of evil, in order that they may more fully indulge in some other, in which is their corrupt taste; or condemn little errors in others that they may escape notice of their own greater breaches of the law of loving kindness to all creatures. (See also Lection lxviii., 18, and lxix., 5.) On which otherwise obscure passages the above remarks may throw a needed light. In these the conditions of the individual are referred to, the one throwing a light on the other. For as with the Church so with each individual composing it, they must progress to

this perfection of character—the "measure of the stature of the fulness of Christ."

The letters beneath were found on an ancient English stone, cube-shaped, "six squared," in a church in Warwick, with these lines, "I, Thou, He, She, We, Ye, It, They; All are one in Me"—supposed to be the work of a Rosicrucian Pantheist of old time.

LECTION LIV. 14-16.—The meaning of these words and this action is very obscure, but if we describe the magic square of 7, it seems to make it intelligible as the mystic symbol of him who regarded everything by number and by measure, and which seems to have reference to the period of his mortal life, 49 years, as well as the number of the Council, Cardinals and Priests of the Church universal, 48, presided over by its Head, 49, which the action of Iesus seemed to symbolize, and in a way, foreshadow.

LECTION LIV. 17-20.—Here we have the original words of Christ, from which Paul adopted his simile in Rom. xii., and in 1 Cor. xii.

MAGIC SQUARE OF SEVEN.
SIGILLUM XTI.

22	47	16	41	10	35	4
5	23	48	17	42	11	29
30	6	24	49	18	36	12
13	31	7	25	43	19	37
38	14	32	1	26	44	20
21	39	8	33	2	27	45
46	15	40	9	34	3	28

21 Summations of 175. Sum total of terms, 225.

This most ancient seal of XT—the High Priest, the At-one-er and At-one-ment of all things, the Lord our Lady Iesu-Maria—is one of the most ancient and sacred symbols of the Xn religion, extending back to pre-Xn antiquity, and symbolising the restoration of the primeval order to the universe, bringing order out of confusion. In this most curious and mystical combination of

text continued on next page

numbers we have 21 summations of 175 (this number cabalistically added being symbolic of XT amid the XII.), viz., 7 rows; 7 columns; 2 diagonals; 1 central cross, which sums 2 ways (5 vertical with 2 transverse numbers, or 5 transverse with 2 vertical); 3 squares made of the 4 corner numbers of the inner, middle and outer courts with 3 vertical, or 3 transverse central numbers. There are other curious combinations, e.g., the sum of the four corner figures of each of the squares being added give 100. Likewise the extreme end numbers of the central cross, while the complementary pairs, wherever found, being added, sum 50. Regarding the symbolism, each number denotes some place or office in the Council of the Church, whose symbol is 120, under her head or his vicar. 49 symbolizes the number of the priesthood of the Universal Church under its head, and 25 that of the Local Church under its head, while 1 represents the Unity underlying all.

It will be noted that 49 symbolizes also the number of years in life of XT by the testimony of His contemporaries, the Jews, and S. Irenæus, B.M., A.D. 189, who received it from the surviving apostles and disciples of the Lord. His ministry having lasted 18 years from the date of His baptism, when He was manifested as the XT, the beloved Son of God.

I.H.W.Y.S.T.

LECTION LV. 1.—This beautiful parable has been sadly mangled in the A.V., and shorn of the opening incident which led to the discourse.

LECTION LVI.—This touching account of the raising of Lazarus is here given as it took place. The verses 13-16 in the Authorised Version are an evident interpolation to magnify the occasion, for, being omitted, the narrative is unbroken and complete without them. As with the daughter of Jairus, so with Lazarus, he was carried to his burial in a state of trance, indistinguishable from death, and by his friends believed to be dead. At the present time in countries where there are mortuaries or waiting rooms for the dead, it is found that five per thousand recover on their way to burial who otherwise would have been buried alive.

LECTION LVII. 4.—The doctrine of guardian angels receives full support from these words. But the Churches of the so-called Reformation have flung away this consoling and helpful belief, with other doctrines of the Christian Church in all ages, the truth of which science and occultism are now showing.

LECTION LVIII. 2.—The charity and comprehensiveness of the true doctrine of Iesus here manifest themselves. It is not a mere narrow creed or belief, but true repentance which merits the forgiveness of God.

LECTION LIX. 11-12.—The teaching of our Lord as to cycles, and the unity of life, in many existences, has been suppressed for long ages, but now sees the light, at the end of the cycle.

LECTION LX. 16.—The same Zaccharias who is mentioned in the beginning as the father of Iohn the Baptist (see Note 111-2), also the Proto Evangelium attributed to Iames, the Bishop of Jerusalem.

LECTION LXI.—All through this chapter the language is highly symbolical, but will present little difficulty to the initiated. v. 12. (See Note in the original "Genesis." Edited by same Editor).

LECTION LXII. 1.—This parable of the ten virgins most accurately indicates the oblivion and indifference which shall come on Christians in the last days of the Christian Church —the days of Laodicean indifference, the Seventh or last age.

LECTION LXIII. 8.—These words are one of the "last sayings" of Iesus, and vividly describe the duty of a Christian Council, so oft neglected.

LECTION LXIV.—The occult teaching, in this discourse, of Iesus to his twelve has been handed down in spirit through the ages, but the world is blind and perceives not. See the same teaching in "New Light on Old Truths," founded on this Scripture.

LECTION LXIV. 3.—Beneath this profound saying of the Ghost Physician, the student cannot fail to notice the intimate and correct knowledge of the human frame, underlying the spiritual truth, which he enunciated. This knowledge has been claimed by science only some centuries later. The innerself-"alternate sex," in every man and woman, which occasionally manifests itself in the dream state, seems to be no mystery to him. (See G. Leland on "The Alternative Sex"; Welby, London).

LECTION LXV. 2.—It has been supposed by some, and not without some reason from the words of the Gospel, that envy and jealousy, and not greed of money, were the cause of Iudas' treachery, because he desired Mary Magdalene, and she had given all her love and devotion to her Master. This inner feeling seems to be concealed beneath the cloak of zeal for the poor. "From that hour he sought to betray him." It is as probable that all three motives urged him, as they do the multitudes nowadays, who grudge magnificence of architecture, music, etc., under the cloak of "utility." "These things ought ye to have done and not left the other undone." But such show their hypocrisy by their reckless contributions to war, and to all manner of pleasures and amusements and luxuries which minister to self. By the spirit of Iudas Iscariot are all such led and dominated.

See next page for Jain's coloured version of a Magic Square of 7x7

A Magic Square of 7x7, back cover,
showing the pattern done at 0 degrees + 90 degrees, that is, 2 times.

The Book of Magic Squares
Volume 1

originally published in 1990 and known as:

The Book of Hamony Squares

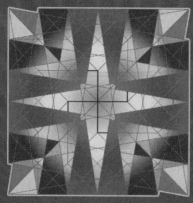

- This book is designed as a Step - By - Step - Join - The - Dots - Draw - Your - Own - Magic - Square - Mandala - Color - In-Activity - Book - Instruction Book for Children & Adults. The Parent or Elder uses it as a Teacher's Manual to direct the younger child to create their own **Crystalline Creations**.
The Magic Square of 3x3 was known as The **Lo-Shu** in ancient China. Printed historic records show it is the oldest known mathematics on the planet circa 5,000 years ago.

- It is a book whose spinal idea is that Mathematics can be **Fun**. Dyslexic children enjoy creating these patterns, as do all children, based on the understanding that children learn mainly via visual imput. This is "**right-feminine-brain mathematics**" at its best.

- It is presented as a non-religious or **omni-sectarian** publication, in that all the anointed Masters & Mistresses of all times are acknowledged and blessed equally.

- One purpose of this book is to cross-pollinate cultural & educational groups, to build bridges rather than barriers, thru the unique symbol of the magic square whose sums of columns, rows & diagonals are **equal**. Occasional references are made to promote awareness to Environmental Harmony or **Ecosophical** causes.

- **Magic Square Art**, also known as Harmony Square Art is a supreme triumph of colour & **psycho-active patterns** (known here as Yantra); many of its timeless qualities have been historically underrated in the West. Although this ancient topic is remote from European traditions, there is a renaissance happening, that has inspired Jain to prepare this book of Mathemagics to meet the creative demands of primary & high-school children & adults. He has successfully taught this subject at international conferences and introduced it to many schools.

- Upon tiling or **tessellation** of the Lo-Shu pattern, the resultant pattern is the **atomic structure of diamond**. This is the principle of **translating Number Into Atomic Art** aka The **Art of Number** permitting blueprints of creation to be revealed, the invisible made visible.

- "With a simple knowledge of Magic Squareology, I have re-educated myself, I love mathematics & its **internal symmetry** when it is transformed into Art. Such is my Bhakti or Devotion & Joy in teaching & radiating It to all Children". **Jain**

jainmathemagics.com

The Gospel Of The Holy 12

Magic Square of 7x7 from Jain's first book:

"The Book Of Magic Squares" volume 1, originally called "The Book Of Harmony Squares".

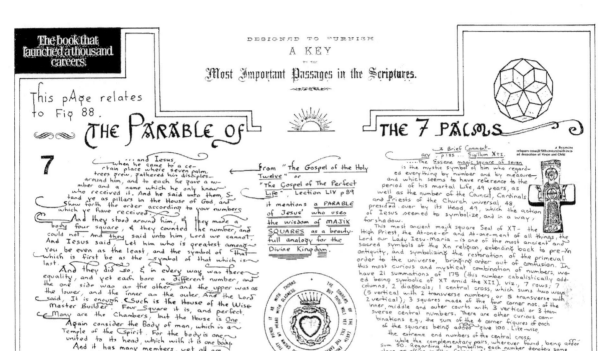

The Book Of Magic Squares, by Jain, Volume 1

Another version of the same Parable of the 7 Palms:

Calligraphy and Art from Jain, extracted from: "The Book Of Magic Squares" volume 3.

The Gospel Of The Holy 12

Rotating the Magic Square of 7x7 Eight Times:

taken from Jain's first book:

"The Book Of Magic Squares" volume 1, originally called "The Book Of Harmony Squares". All geometries are hand-drawn.

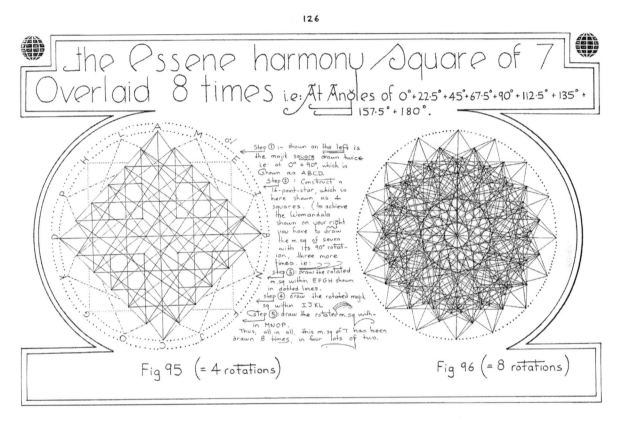

The Book Of Magic Squares, by Jain, Volume 1

The Magic Square of 11x11, the Seal of Christ.
(**Ecclesiae Militantis Sigilum**)
This is a typical page of two columns per page, in small font, still readable, but preserved in its original 1974 presentation to show that nothing has been altered.

188 A BRIEF COMMENTARY.

LECTION LXVIII.—Again, studying these "dark sayings" so difficult to understand, recourse has been had to certain figures known to the early gnostics. (*See the "Squares and Circles" by the Editor of this Gospel.*) The Magic Square of 11 has been found wonderfully to explicate, symbolically at least, to the mystical, the meaning of the passage. The form exactly illustrates what the Lord in symbol taught to his disciples of the bringing forth of order out of disorder, perfection out of imperfection, and out of deficiency fulness. Compare the Magic Square given below with the natural square which any one can form by writing the numbers in consecutive order; the result is at once seen, and may help to arrive at the meaning of this very mystical passage :—

MAGIC SQUARE OF ELEVEN PERFECTED.
(*Ecclesiæ Militantis Sigillum.*)

56	117	46	107	36	97	26	87	16	77	6
7	57	118	47	108	37	98	27	88	17	67
68	8	58	119	48	109	38	99	28	78	18
19	69	9	59	120	49	110	39	89	29	79
80	20	70	10	60	121	50	100	40	90	30
31	81	21	71	11	61	111	51	101	41	91
92	32	82	22	72	1	62	112	52	102	42
43	93	33	83	12	73	2	63	113	53	103
104	44	94	23	84	13	74	3	64	114	54
55	105	34	95	24	85	14	75	4	65	115
116	45	106	35	96	25	86	15	76	5	66

33 Summations of 671.   Sum total of terms, 7,381.

LECTION LXIX. 8, 9.—This saying of Jesus is very difficult to the popular mind, as apparently reversing the original injunction in Gen. 1-8, "Be ye fruitful and multiply." To understand this, it must be borne in mind that the promise of a Messiah to redeem the world, has, from the earliest

LECTION LXVIII.—Again, studying these "dark sayings" so difficult to understand, recourse has been had to certain figures known to the early gnostics. (See the "Squares and Circles" by the Editor of this Gospel.) The Magic Square of 11 has been found wonderfully to explicate, symbolically at least, to the mystical, the meaning of the passage. The form exactly illustrates what the Lord in symbol taught to his disciples of the bringing forth of order out of disorder, perfection out of imperfection, and out of deficiency fulness. Compare the Magic Square given below with the natural square which any one can form by writing the numbers in consecutive order; the result is at once seen, and may help to arrive at the meaning of this very mystical passage:—

MAGIC SQUARE OF ELEVEN PERFECTED.
(Ecclesiæ Militantis Sigillum.)

56	117	46	107	36	97	26	87	16	77	6
7	57	118	47	108	37	98	27	88	17	67
68	8	58	119	48	109	38	99	28	78	18
19	69	9	59	120	49	110	39	89	29	79
80	20	70	10	60	121	50	100	40	90	30
31	81	21	71	11	61	111	51	101	41	91
92	32	82	22	72	1	62	112	52	102	42
43	93	33	83	12	73	2	63	113	53	103
104	44	94	23	84	13	74	3	64	114	54
55	105	34	95	24	85	14	75	4	65	115
116	45	106	35	96	25	86	15	76	5	66

33 Summations of 671. Sum total of terms, 7,381.

LECTION LXIX. 8, 9.—This saying of Iesus is very difficult to the popular mind, as apparently reversing the original injunction in Gen. 1-8, "Be ye fruitful and multiply." To understand this, it must be borne in mind that the promise of a Messiah to redeem the world, has, from the earliest times, begotten in the woman of the Hebrew nation, to whom it was specially given, that insatiable desire for offspring, each woman thinking of herself as the possible mother of Him who was to come and save. Iesus, the true Prophet, seeing the tendency to the propagation of the unfit (as we now see) to bring want, misery, squalor, vice and crime, through the inability of most parents to bring them up as they should be, owing to the curse of competition and greed, here proclaimed to Salome, in answer to her query, that he would reverse all this tendency among his followers, and through them, extend this reversal to mankind at large, for He, the desire and hope of nations, having come, there was no longer any supposed necessity or reason for such increase, of which He, the Prophet of God, fully foresaw the evil, in the ages to come, as we now fully experience it.

LECTION LXX. 1-5.—Long has Iesus suffered reproach, for those words so falsely attributed to him, in place of the impulsive Peter, who spoke them, and with whose character they were in full harmony.

LECTION LXXI. 1-4.—Twice the Lord is said to have performed this symbolic act. Surely, at his return, it will be his first work! For since the first ages till now the spirit of the world ruleth, and mammon is dominant, and every kind of wickedness in the name of religion, zeal for purity, etc.

LECTION LXXII. 1.—In the language of the Churches of this day, there is but one mansion in the Father's house, and *that* is claimed by each of over 300 different sects as its own, and all outside are damned, not for their evil deeds, but because they cannot see as their rulers profess to see!

LECTION LXXIII. 1-6.—"I am the true Vine, ye are the branches"—in unity with the stem by the continual possession of the One Life, not by *mere* external unity, valuable as this is, and certainly not by a dead uniformity of opinion in all things. "*Tot homines tot sententiæ.*"

LECTION LXXV. 1.—Iacob is the same as Iames—called "the great." Nathanael is Bartholomew. There is no proof that Iude was the same with Thaddeus, as is alleged by some. The number at first seems to have been twelve exclusive, or thirteen (to the world's eye) including Iudas Iscariot, till he should manifest his falsity by his treachery, when he went out directly before the holy supper, leaving Iesus with the twelve—the complete number of Apostleship, which, being even, admitted of no one among them being "Master," save Iesus, who was over them.

v. 2.—Whether the appearance of the Master and his disciples in symbolic festal garb may not have been seen only by the spiritual eye of some of the disciples or not, the lesson is the same. Reverence and love of beauty and order are to be seen in God's House—symbols of the glorious garments of that Being Who is the Eternal Mystery and Beauty manifest in all things.

vv. 15, 16.—That Twelve is the complete number of the Apostleship and that Iesus sat down "*with his twelve*" at the holy supper before his crucifixion, seems evident from the received gospels, and still more so, from the fragments lately brought to light. Iudas Iscariot appears then to have been *among* the twelve but not of them, therefore before the Eucharistic rite is celebrated "he goes out." If there were any ill omen at all about the number thirteen it would therefore be thirteen as the number of Apostles present, exclusive of the Master and Head. But to thirteen, inclusive of the presiding host, no ill omen could attach, but the reverse.

LECTION LXXVI. 4.—There are two other alternative versions of these circumstances of the last supper in the A.V.—First, that of St. Iohn who, in the received version, expressly affirms that Iesus was crucified on the very day of the Passover and consequently the Eucharist was instituted the day before and not on the feast day itself, and the Passover was on the morrow after the trial on the day of the crucifixion. Secondly, that of the three other gospels, which all affirm that the Eucharist was instituted on the Passover the pascal lamb was slain. If the latter, it must be remembered that the Essenes (of whom Iesus was apparently one), were by Iewish regulation allowed a separate table at which no lamb or other flesh-meat was eaten, as they were vowed abstainers from blood sacrifices and the eating of flesh. If the former it was not the Passover at all, and Iesus was not bound as a Iew to eat of a lamb. In neither of these cases, therefore, was Iesus under the alleged necessity of killing a lamb and eating of flesh-meat in order to fulfil the law. In any case the causing of an innocent lamb to be killed and the eating of such is contrary to all that is known of the character of Iesus the Christ, whose tender love extends to all creatures. If Iesus was not an Essene, then nothing can be said against the accuracy of this version of the holy supper, and the charges brought against him in the account of the trial as now given by the Spirit.

v. 9.—"Bread," *i.e.* unleavened cakes of pure meal such as in use at the Passover. "Wine," here and through the Gospels, as used by Iesus and His disciples, means "the fruit of the Vine," which is pure wine mingled with four or two parts of pure water, the latter mystically representing the humanity, and the former the Divine Spirit. The strong fermented wine of modern use was never used on such festive occasions, nor even generally, except thus mingled with water. It is to be noted that the Saviour consecrated the Eucharist by invocation of the Holy Spirit, and this has been faithfully followed by all Churches of the East, the words of institution being merely recited before, as a historical preamble, giving the authority for the action, and in no case as the words of consecration, according to the corrupt use of the West.

v. 13.—In the received Gospel Iudas is consigned to eternal perdition, but it appears rather that he who was all compassion and prayed for his murderers, prayed also for the man who was overmastered by his passions, blinded by envy, jealousy, greed of money, or, as some say, by desire to push matters to their conclusion, and procure some decisive miracle that would establish the claim of his Master to set up a temporal kingdom.

v. 26.—It is not stated whether there was any musical accompaniment, as is usual in the religious dances and processions of the East, but if so it was probably of the simplest, such as the Pipe, used on such occasions.

v. 27.—The Mazza, or unleavened cake, to which may the word "Mass" be traced as applied to the Eucharist, or "Breaking of Bread"—but preferable perhaps is the interpretation of "ite missa est"—the oblation (= prayer) is gone, "*sent up.*"

v. 30.—Here was perhaps more probably the sole motive actuating Iudas—his ambition—the desire to see a miracle, and the early sovereignty set up before the time.

LECTION LXXVII. 2.—Here the Lord addresses Simon, not Peter. In the A.V. confusion has arisen owing to the same name being given to two Apostles, and Peter is made to reply. It does not seem likely that one who thrice betrayed the Lord should by him have been placed in the highest authority, as it subsequently appears that Peter was.

LECTION LXXVIII. 12-18.—The belief that Peter denied his Master is probably owing to two of the Apostles bearing the same name, Simon Peter and Simon the Canaanite. Here we are given the right version. The error is one that might have been easily made. It is worthy of notice, that this ancient Gospel attributes to Simon (not to Simon Peter) the thrice denial of Iesus, and this fully exonerates Peter from the baseness generally attributed to him, and to which there is no allusion in his writings, but rather the reverse in the accepted gospel, where he was first to draw the sword in defence of his Master.

A TRUE LIKENESS OF
OUR SAVIOUR.

The true likeness of Jesus.

THE true Likeness of Our Saviour (on opposite page) was copied from the portrait carved on an emerald by order of Tiberius Cæsar, which emerald the Emperor of the Turks afterwards gave out of the Treasury of Constantinople to Pope Innocent VIII. for the redemption of his brother, taken captive by the Christians.

The following was taken from a manuscript now in the possession of Lord Kelly, and in his library, and was copied from an original letter of Publius Lentullus at Rome. It being the usual custom of Roman Governors to advertise the Senate and people of such material things as happened in their provinces in the days of Tiberius Cæsar, Publius Lentullus, President of Judea, wrote the following epistle to the Senate concerning our Saviour:—

"There appeared in these our days a man, of the Jewish Nation, of great virtue, named JESUS CHRIST, who is yet living among us, and of the Gentiles is accepted for a Prophet of truth, but His own disciples call Him the *Son of God*—He raiseth the dead and cureth all manner of diseases. A man of stature somewhat tall, and comely, with very reverent countenance, such as the beholders may both love and fear, his hair of (the colour of) the chestnut, full ripe, plain to His ears, whence downwards it is more orient and curling and wavering about His shoulders. In the midst of His head is a seam or partition in His hair, after the manner of the Nazarites. His forehead plain and very delicate; His face without spot or wrinkle, beautified with a lovely red; His nose and mouth so formed as nothing can be reprehended; His beard thickish, in colour like His hair, not very long, but forked; His look innocent and mature; His eyes grey, clear, and quick. In reproving hypocrisy He is terrible; in admonishing, courteous and fair spoken; pleasant in conversation, mixed with gravity. It cannot be remembered that any have seen Him *Laugh*, but many have seen Him *Weep*. In proportion of body, most excellent; His hands and arms delicate to behold. In speaking, very temperate, modest, and wise. A man, for His singular beauty, surpassing the children of men."

GROUND PLAN OF A CHRISTIAN CHURCH
ARRANGED FOR
APOSTOLIC WORSHIP AND DISCIPLINE.

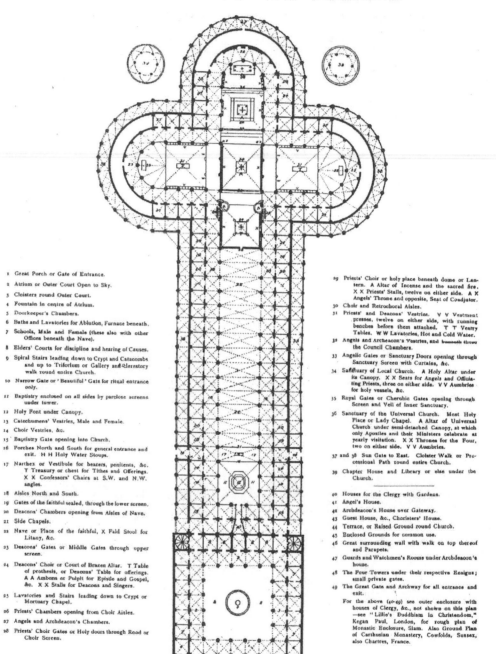

1. Great Porch or Gate of Entrance.
2. Atrium or Outer Court Open to Sky.
3. Cloisters round Outer Court.
4. Fountain in centre of Atrium.
5. Doorkeeper's Chambers.
6. Baths and Lavatories for Ablution, Furnace beneath.
7. Schools, Male and Female (these also with other Offices beneath the Nave).
8. Elders' Courts for discipline and hearing of Causes.
9. Spiral Stairs leading down to Crypt and Catacombs and up to Triforium or Gallery and Clerestory walk round entire Church.
10. Narrow Gate or 'Beautiful' Gate for ritual entrance only.
11. Baptistry enclosed on all sides by parclose screens under tower.
12. Holy Font under Canopy.
13. Catechumens' Vestries, Male and Female.
14. Choir Vestries, &c.
15. Baptistry Gate opening into Church.
16. Porches North and South for general entrance and exit. H H Holy Water Stoups.
17. Narthex or Vestibule for bearers, penitents, &c. T Treasury or chest for Tithes and Offerings. X X Confessors' Chairs at S.W. and N.W. angles.
18. Aisles North and South.
19. Gates of the faithful sealed, through the lower screen.
20. Deacons' Chambers opening from Aisles of Nave.
21. Side Chapels.
22. Nave or Place of the faithful, X Fald Stool for Litany, &c.
23. Deacons' Gates or Middle Gates through upper screen.
24. Deacons' Choir or Court of Brazen Altar. T Table of prothesis, or Deacons' Table for offerings. A A Ambons or Pulpit for Epistle and Gospel, &c. X X Stalls for Deacons and Singers.
25. Lavatories and Stairs leading down to Crypt or Mortuary Chapel.
26. Priests' Chambers opening from Choir Aisles.
27. Angels and Archdeacon's Chambers.
28. Priests' Choir Gates or Holy doors through Rood or Choir Screen.
29. Priests' Choir or holy place beneath dome or Lantern. A Altar of Incense and the sacred fire. X X Priests' Stalls, twelve on either side. A X Angels' Throne and opposite, Seat of Coadjutor.
30. Choir and Retrochoral Aisles.
31. Priests' and Deacons' Vestries. V V Vestment presses, twelve on either side, with running benches before them attached. T T Vestry Tables. W W Lavatories, Hot and Cold Water.
32. Angels and Archdeacon's Vestries, and beneath these the Council Chambers.
33. Angelic Gates or Sanctuary Doors opening through Sanctuary Screen with Curtains, &c.
34. Sanctuary of Local Church. A Holy Altar under its Canopy. X X Seats for Angels and Officiating Priests, three on either side. V V Aumbries for holy vessels, &c.
35. Royal Gates or Cherubic Gates opening through Screen and Veil of Inner Sanctuary.
36. Sanctuary of the Universal Church. Most Holy Place or Lady Chapel. A Altar of Universal Church under semi-detached Canopy, at which only Apostles and their Ministers celebrate at yearly visitation. X X Thrones for the Four, two on either side. V V Aumbries.
37 and 38. Sun Gate to East. Cloister Walk or Processional Path round entire Church.
39. Chapter House and Library or else under the Church.
40. Houses for the Clergy with Gardens.
41. Angel's House.
42. Archdeacon's House over Gateway.
43. Guest House, &c., Choristers' House.
44. Terrace, or Raised Ground round Church.
45. Enclosed Grounds for common use.
46. Great surrounding wall with walk on top thereof and Parapets.
47. Guards and Watchmen's Rooms under Archdeacon's house.
48. The Four Towers under their respective Ensigns; small private gates.
49. The Great Gate and Archway for all entrance and exit.

For the above (40-49) see outer enclosure with houses of Clergy, &c., not shown on this plan —see "Lillie's Buddhism in Christendom," Kegan Paul, London, for rough plan of Monastic Enclosure, Siam. Also Ground Plan of Carthusian Monastery, Cowfolds, Sussex, also Chartres, France.

A Ground Plan of the first Christian Church, arranged for Apostolic Worship and Discipline. The upper room in Jerusalem (presided over by James, its first Angel or Bishop, "brother of our Lord") being the "nucleus" (represented by the chancel), to which are added the features of Christian Cathedral Churches of some various centuries later, so far as they harmonize with the original Idea.

Reduced from an ideal plan in the Editor's possession.

The Christian Essene Church Floor-Plan

Meditation.

O Central Fount of life and love,
From Thee we come, to Thee return
Through many wanderings, far and near
At last we in Thy Home appear

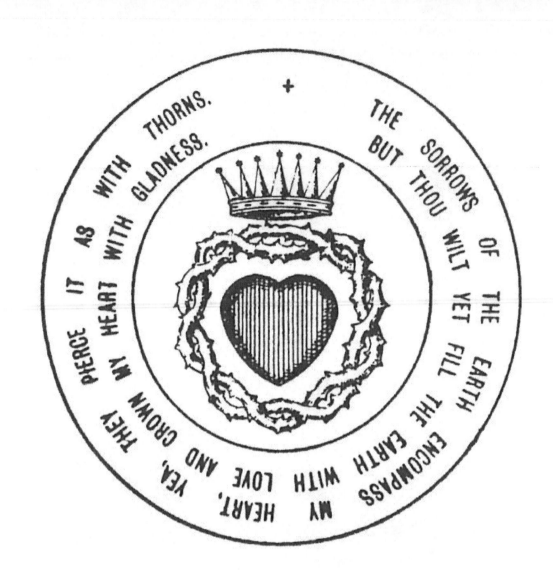

from the dead: and the Life everlasting of the just. The Ages of Ages: and Rest in God for ever.—Amun.

24 And as the smoke of the incense arose, there was heard the sound as of many bells, and a multitude of the heavenly host praising God and saying:

25 Glory, honour, praise and worship be to God: the Father, Spouse, and Son: One with the Mother, Bride and Maid: From Whom proceedeth the Eternal Spirit: By whom are all created things. From the Ages of Ages, Now: and to the Ages of Ages—Amun—Alleluia. Alleluia, Alleluia.

26 And if any man take from, or add, to the words of this Gospel, or hide, as under a bushel, the light thereof, which is given by the Spirit through us, the twelve witnesses chosen of God, for the enlightenment of the world unto salvation: Let him be Anathema Maranatha, until the coming of Christ Iesu-Maria, our Saviour, with all the Holy Saints. Amen.

Here endeth the Holy Gospel of the Perfect Life of Iesu-Maria, the Christ, the Son of David after the Flesh, the Son of God after the Spirit. Glory be to God by Whose power and help it has been written.

THE EPISTLE OF APPOLLOS THE PROPHET.

APPOLLOS TO HIERASTHENES, GREETING.—

1 TOUCHING the matter whereby thou didst enquire in thy last epistle, I will inform thee even as I have received. I Apollos was in my house in Nazareth after the Holy City had been taken by the Romans, and the Temple of God destroyed, even as the Lord had told us.

2 And as the sun went down and I was resting from my work, the room was filled with a bright light and there appeared unto me Agella, my sister (who had been reported as dead with many others of the brethren who were in the Holy City at the time of the siege and who have never since been seen by any to this day).

3 And Agella spake to me saying, BROTHER, why grievest thou for me, and for the fall of Ierusalem and for the Holy House. Grieve rather that thou wast left behind when we with others of the brethren who were ready were taken up from the earth.

4 For when the city was sorely besieged and the battle was the most fierce and the confusion great and terrible, there was seen by all a great wonder in the heavens.

5 For the Lord himself appeared from the clouds with her to whom he first appeared after he rose from the dead, who announced his resurrection to the twelve, and the holy angels, according to the word that he spake unto us while he was in the flesh.

6 And we who sore longed for deliverance and were ready for his appearance were caught up to him in the clouds with Iohn, who alone of the twelve remained (whether in the body or out of the body I knew not).

7 It was in a moment and we were changed in the twinkling of an eye, and those who were his enemies saw it and fled in great confusion and fell on the swords of the Romans and perished, and to me alone has it been given that I should appear unto thee for thy comfort my brother, and for the consolation of those that are left behind and those that shall come after them, that they may believe in the words spoken by the Lord before he suffered.

8 Farewell brother, and go and comfort those that are left, for there will arise those who will deny that he

166 THE EPISTLE OF APPOLLOS TO HIERASTHENES.

returned as he said, because none of those who saw his appearance are left behind to witness thereof.

9 But believe thou that the Christ shall return again at the end of the Age in glory.

10 AND I arose and went to some of the brethren and told them these things, but they seemed to them as an idle tale, for they answered, If thy sister and the others were taken, why have we been left behind in the misery of this world? Surely they have fallen by the sword also, and it was a vision, and we which are left behind shall perish likewise?

11 And I returned to my home and held my peace, for I was in doubt, and said, If the thing is true it will be brought to light in a future day, for the Lord certainly did say that " before this generation should pass away all these things should be," even as my sister hath told me they have been.

12 They who are with me salute thee. Peace be with thee, and to all in thine house.

EPISTLE OF APPOLLOS TO HIERASTHENES.

DATE ABOUT 71 AND 72.

NOTE.—This fragment must be received or rejected on its own interial evidence, as compared with Matt. xxiv. 29-34. For either this event did not happen as related, and the Lord Iesus was deceived and prophesied falsely, or else the Lord's words were fulfilled as he said " before this generation passes away," and this ancient letter is a reference to it, the only reference which, by the nature of the case, we can have, unless others of those "taken up" were now to appear and add their testimony.

A BRIEF COMMENTARY:

BEING NOTES ON

THE GOSPEL OF THE TWELVE;

KNOWN ALSO AS

The Gospel of the Perfect Life.

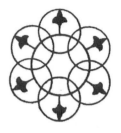

of the Early Christian Church (reaching back, perhaps, in some truths, to the days before Noah), which we see glimpses of elsewhere in the writings that remain to us. The decree of the Church in her first general council at Ierusalem, founded on it as a deep psychological insight, and given under the direct influence of the Holy Spirit, is well known, though generally forgotten or ignored by Christians; for the logical deduction is total abstinence from flesh-meat, which can only be obtained for food by the destruction of innocent life, whether by bloodshed or by strangling. And such destruction of life for selfish ends is placed with fornication and other deadly sins. Some portions of the Epistles of S. Paul in the A.V., notwithstanding all these, are manifestly interpolations of a later date after the discipline of the Church had been relaxed, and the evil customs and fashions of the world had sought and gained admission in the Church in union with a corrupt state. Both this symbol of Faith and the 12 Precepts of Iesus have been published before in a separate leaflet form in the year 1881.

v. 22.—It has been asserted by Church historians of all shades of opinion that incense was not used in the Christian Church till the fourth century. Here we have evidence of the contrary, and proof that in the early Church of Ierusalem it was in use at the time this Gospel was written. This with other ceremonies, ornaments, etc., not peculiar to Iudaism (e.g. the high priest's golden plate or mitre, which S. Iames is said to have worn as president of the Church of Ierusalem), were retained. The truth seems to be that the use of it was laid aside for a time as a matter of expediency on account of the danger to the lives of the brethren, as it helped materially their persecutors to find out their hidden places of meeting. When the persecutions were over and the Church emerged from the concealment of the catacombs into the light of open day in the fourth century, then it was resumed, and this was the only reason of what could exist, for its discontinuance being of divine appointment as we have seen, though not essentially necessary any more perhaps than music, or lights or vestments.

v. 22.—Probably as a traditional memory of this recorded event, a peculiar custom exists in the East to this day. The Great Bell of the Kremlin Tower is heard during the entire time of the chanting of the Creed. The same in other Oriental Churches, where every bell in the Church rings during the creed, as well as at the elevation and other parts of the Liturgy. (See O'Brien's "History of the Mass," &c.)

Again, referring to the Essenes as a religious body, the popular impression is that the Cenobite or Monastic life did not rise in Christendom till the Fourth Century. The fact is, it was coeval with, if not antecedent to, Christendom, as among the Essenes and Therapeutæ, and even before them in the "school of the Prophets" in the Iewish Church. The Carthusian Monasteries in the Catholic Church give a good idea of these early monks and nuns, and their mode of life, &c. (See Arthur Lillies "Christianity and Buddhism," larger work). It is the natural outcome of earnestness and devotion, despised and rejected by the world that *will not* receive nor give heed to higher teachings but only to its own self-interests, insanities and follies, and reject the life of obedience to God's laws. "Ye are the salt of the earth, a city set on a hill, a light shining in a dark place." Thus, with few exceptions, were these institutions of old, and still are, where the evil influence of the world has not crept in like a serpent coiling round the Tree of Life.

v. 26.—Here we subjoin the ancient Anathema omitted in the first Edition which we are now admonished to restore to the complete Edition as now published.

COMMENTS of the EDITORS

The Gospel of the Holy Twelve
Comments of the Editors
Transcribed and compiled
by Rev. Mark Wilcox D.D.

Section 1, Lections 1 Thru 10

LECTION 1.1 The opening paragraph of this Gospel was evidently before the eyes, or in the mind of St. Luke when he wrote his Gospel. (see Luke 1:5). This is only one of several instances where this Gospel, or the words of Iesus recorded in it, are used subsequently, without specially indicating the fact (as shewn further on), being well-known to his hearers at that time.

LECTION 2. 10.- "Joseph begat (of Mary the Virgin, his wife) Iesus, who is called the Christ."- Curetonian and Lewis's Syriac, MS. ; and several of the oldest Latin MSS., in Matt. I.16, A.V.

LECTION 2. 21-25.-The canticle of Ioseph here given is very similar to a certain portion of the book of Isaiah; indeed, appears to be taken from it, as Iohn borrowed from the Old Testament prophets. It has been omitted in all other Gospels extant. It is of singular beauty, and appropriate for use at Matins, as *Magnificat* is for Vesper, the Song of Zacharias finding an equally appropriate place at Nocturns.

LECTION 4. 1 -The accepted date of the birth of Christ as corrected in the A. V. is A.M. 4000, or A.D. 1. This being so, his second visit to the Temple A.M. 4012, and after that his travels about A.M. 4018-4030; his Baptism A.M. 4031 ; His Transfiguration on the Mount, 4042 ; and his Crucifixion A.M. 4049, leaving eighteen years for his public ministry ; and his numerous teachings, which S. Iohn declares would fill a vast number of books, more than could be contained (comprehended by the world).

LECTION 4. 4 - The animals here mentioned are sacred to the Deity in various countries and religions, the Cat and the Dove being specially honored and protected in Egypt (the most ancient centre of civilization, religion, philosophy and true science), as the symbols of Isis, the foreshadower of the "Divine Mother" of Christianity. Egypt (with her Trinity of Father, Mother, Child) gave refuge and sanctuary to the Infant Christ, Who came forth from thence to redeem humanity. The cat is not wilfully a "cruel animal," as falsely alleged by the ignorant, no more than the babe which torments it in ignorance of the pain it gives. Far more cruel are human beings, who torture and destroy millions of innocent creatures to gratify a depraved appetite or to minister to their vanity, or their lust for cruel experiment. The cat truly, as alleged by occultists, both ancient and modern, "the most human of all animals," and it is probable it was for this reason that it appears as the favourite animal of Jesus who was ever the friend of the despised, maligned and neglected although the most loving, gentle and graceful of all animals, rather than the more self assertive dog, especially as taught by man to hunt and to worry.

LECTION 4. 12.-Iesu Maria is the complete name. Jesus, he shall save, Maria, his people. Jesus is only the first part of the Holy Name, He saves His people, not at once, the entire human race, but those of

The Gospel Of The Holy 12

170

goodwill- *homines bonce voluntatis* - men and women of peace, and obedient to the divine law; and by these, their brethren through the ages, who *will* to be saved. The first part of the sacred Name seems to be generally used in the Gospel, as indicating that only the first part of his mission is *now*. When all men and women are gathered in, then will Christ be manifest as the complete Saviour, Iesu-Maria.

LECTION 5. 9.-Note the beautiful lesson taught by these words. They look in vain for the signs of God who forget the needs of the poorer brethren and their beasts under their care. To look upon the needs of these who cannot speak (in human tongue) is to find the bright light they lose who only look upwards.

LECTION 5. 16. -Alluding to 2 Chron. xxiv. 20, in the Ierusalem Talmud, and also in the Babylonish, is an account of a priest named **Zacharias**, who was slain in the court of the priests near the altar, and whose blood never ceased to bubble from the earth, till a great number of priests and rabbins were slaughtered *(Talmud Hierosal,* fol. 69).

In the Protevangelium attributed to Iames, the first Bishop or Angel of the Church in Ierusalem is introduced the present story of Zacharias, and that Herod who slew the infants in Bethlehem slew also Zacharias the priest in the Temple when he said that he knew not where his infant son Iohn was hidden. It is this story, and not the incident in Chronicles, that most probably is referred to in a latter part of the Gospel by Iesus, being fresh in the memories of that generation, and so more likely to fasten attention.

LECTION 6. 5. -In what way this prediction is to be fulfilled is not as yet made manifest - whether Iesus shall yet be manifest and received by his people as the Two-in-One, the All-gentle as well as the All-powerful, or whether He shall assume the feminine form, or whether He shall be manifest with His counterpart. Many false Christs shall come with signs and lying wonders.

LECTION 6. 13. -Iosephus mentions a section of the Essenes, or Iessenes, who, unlike the great majority of them, lived in "honourable marriage," observing their rules and customs in all other matters, such as abstinence from blood sacrifices, flesh eating, etc. Some consider it most probable, therefore, that at this period Iesus married, according to the usual custom of the Iews, and in his case especially, that he might have full experience of human life, and thus be a perfect Example for all, knowing the joys and sorrows of all,-and that it was just before his further travels preparatory to his entrance into the Ministry that he lost by death the wedded partner of his youth. He was " in all things like as we are, yet without sin"

LECTION 7. 4.-The fruit of the Carob tree ("S. Iohn's Bread") ; not the insect of that name, as is supposed by the people in general.

LECTION 7. 10,-As noticed before, the Essenes did not frequent the blood sacrifices of the Temple. Iohn and Iesus acted accordingly.

LECTION 8. 2. -This "bright light" at his baptism is mentioned in the "Gospel of the Hebrews," which is undoubtedly the original Gospel of S. Matthew, and the one used in the primitive Church of Ierusalem, and identical with this.

Iustin Martyn quotes this Gospel as the original Gospel of Matthew, and endeavours to explain away the supposed "heresy" in the words, " This day have I begotten thee," which shows that the present Gospel of Matthew could not have been extant in his time, else he would have quoted it with gladness as omitting these words.

v. 7.-The earthly ministry of Iesus, beginning at thirty years of age, complete and continuing till his death at the age of forty-nine, must therefore have lasted much longer than is generally supposed, even eighteen years. During the latter part of it, the Iews who knew him attested that he was then " not fifty years old."

LECTION 9. 1. -The Essenes or Nazarenes, somewhat like the Indian Yogi, sought to attain divine union by solitary meditation in unfrequented places. In the monastery of our Lord on the summit of Quarantania, a cell is shown with rude frescoes of the event. This mountain is about 18,000 feet high, in a barren and desolate region east of Ierusalem, north of the road to Iericho, overlooking the valley of the Iordan.

v. 2-9.-Observe, the temptations are addressed to the fourfold nature of man, as recognised by the ancient

Egyptians. 1st.-To the outer body, with its physical needs. 2nd. -To the inner body, the seat of the senses and desires. 3rd.-To the soul, the seat of the intellect.

LECTION 9. 3.-In all the ancient initiations woman was one of the temptations placed in the way of the aspirant. That this was not omitted in the trial of the "Perfect Man" we may be certain, and we are expressly told in the Epistle to the Hebrews that "he was in all points tempted even as we are." Why the writers of the Canonical Gospels omitted this trial, or whether it was dropped out of the original by accident we cannot say, but here we have it restored in its place. It is evidently inculcated by Iesus in this second temptation (what has always been known to the wise) that adepts should store up their physical strength for work on a higher plane, and this Iesus did for the work of the ministry as an example for all who would follow him and heal the bodies and souls of others.

Here we have one of the many passages which show that the words attributed to the writers of the Epistles are quotations from this Gospel, and that such portions at least were extant in their time.-e.g., I. John iii. 9. (A. V.).

Section 2 Lection 11 Thru 20

LECTION 11. 1-2.-There are two anointings by Mary Magdalene recorded. The first was to his prophetical ministry, the last preparatory to his self-oblation unto death on the cross in the upper room, and his subsequent murder by the Roman authorities and the Iewish priests.

LECTION 12. 3-4.-Iesus being a Yessene (Essene) could not drink intoxicating wine, and it is to be remarked here, that he did not provide it. He poured water into jars, and they tasted it as wine unfermented, or, if fermented, with four times or least twice its volume of water, which makes what is termed all through the Gospel the *"fruit of the vine."* It is impossible that Iesus could sanction drunkenness, though his enemies slandered him as a "wine-bibber."

v. 16.-Two modes of reckoning time were in use. The Roman, from 12 midnight to 12 midnight. The Iewish from 6 a.m. (mean time is here spoken of) in the even to 6 p.m. of next even. The Iewish hours, adopted from the Temple in the Christian Church in her devotions, were as follows:

6 p.m. 1st watch, Vespers. Ferial.
9 p.m. 2nd watch. Nightfall ("Compline" Lat. use).
12 midnight 3rd watch. Nocturms.
3 a.m. 4th watch. Daybreak. (Lauds).
5-6 p.m. Seventh or last hour of the night.
6 a.m. Matins (or "Prime" Lat. use). First hour.
9 a.m. Terce. Third hour.
12 midday. Sext. Sixth hour.
3 p.m. Nones. Ninth hour.
5-6 p.m. Eleventh or last hour of the day. Vespers. Festal. Really "Compline" in its true sense.
The "seventh hour," in this place is therefore 1 p.m. of our reckoning; whether by Iewish or Roman time- 13th hour in some countries.

LECTION 13. 6.-The effects of his education in Egypt and his travels in other countries and knowledge of their religion and mysteries are here clearly seen in the largeness of the heart of Iesus, and his sympathy with all men. He is the true Catholic, who excludes none from his love whose hearts are unto righteousness, while he pities those that are not, knowing the terrible fate that awaits them.

LECTION 14. 4.-Miracles are not violations of the laws of nature, but rather suspensions of lower by higher laws- wonders wrought by using wisely the subtle forces of nature, (whether by seen or unseen agencies) unknown to the science of the day, and in advance of the knowledge of the people. Many are the spiritual agencies, the knowledge of which we are now recovering, but which have existed and acted all through the ages. Occult phenomena also appear to have been used by religious teachers in all ages in the East, to attract the attention of the listless and thoughtless, and having roused and secured their interest, to teach them spiritual truths or give them higher revelations; just as in the West. in modern times, they Bound a bell, or sing an " invitatory" or a hymn to "call the people."

LECTION 15. 4.-The houses in Palestine were constructed with flat roofs, and entrance was easily made into the court below without entering by the door below.

LECTION 16. 1.-"Levi" is by tradition identified with Matthew, the writer of the second of the four Gospels (as received by the Church), Mark being the first of the four Evangelists, though placed second in the A. V.
v. 7. -It was the custom in Palestine to use the skins of animals to hold wine as we do glass bottles, and such leathern bottles when filled with new wine were liable to burst by reason of the fermentation of the wine within them.

LECTION 17. 3.-Iudas Iscariot is here called a Levite. It may be symbolical of the fact, that the older priesthood was the bitter enemy of Iesus, the Prophets and the Priest of the newer Christian Dispenssotion.
v. 2-5. -Here we have a flood of light thrown on an obscure passage in Ephesians iv. 11, referring to an event of which there is no record whatever in the A. V. or in any other version of the Gospels which has come down to us. Plain enough is the passage in Ephesians as it stands, but obscure in its reference; and the only body of Christians who have in later times restored this ancient fourfold ministry is the "Catholic Apostolic Church," but with this difference, that what Iesus intended to be a permanent order, they have made only a lifetime institution, dying with the men that fill the office, at present the one left being removed by death. Under Jesus, the High Priest, or chief Shepherd and Bishop of the Universal Church, while the two and seventy afterward sent forth were the deacons in the higher ministry, altogether making the full number a hundred and twenty.
v. 6.9. -These words leave no doubt that the organization which Iesus first established was based on the older organization of the Yessenes (similar to that of the Buddhists), and from which have come the monasteries, friaries, and sisterhoods of the Christian Church, which have always been popular with the poor, and befriended them in times of trouble, and set them an example of Godly living; the corruptions and abuses, which set in now and then, being no argument against the use. They were a continual protest against the ways of the world, its vices and luxury and evil pursuits. "Leave all and follow me," was the continual call of the master, to those who could receive it.

LECTION 18. 1.-This number (seventy-two), symbolizring amongst the Jews the Nations of the Earth, and late-denoting the Diaconate of the Church Universal (in priestly orders), was afterwards selected by the Christian Church as the complete number of its cardinals, as it had been before the number of members of the Jewish Sanhedrim.

LECTION 19. 2. -There are two versions given of the Lord's Prayer, this one, the fullest, being given to the Twelve and their companions, and a shorter form afterwards to the people in his Sermon on the Mount.
v. 5, 6.-An ancient saying, long lost to the Church. The all-pervading nature of Deity seems plainly taught, which, in a recently recovered fragment, is obscure.

LECTION 20. 9.-There is no trace of the events between Lections xviii. and xx. other than this giving of the form of Prayer as a model for all time.

Section 3 Lections 21 Thru 30

LECTION XXI. 2-6. -This touching incident is to be found also in a very ancient Coptic fragment of the Life of Iesus- others of a like nature also recorded in their places in this Gospel, show how he, the Divine Saviour of the world, regarded the ill-treatment of the "lower" animals as a grievous sin.

v. 12. -The divine love of Iesus for all God's creatures is everywhere evidenced by this Gospel, and his belief that all life is one, is abundantly justified by the teaching of true modern science, physical and occult.

LECTION XXII. -The daily increasing discoveries in modern times of cases of trance or of suspended animation, in which those carried to burial certified as dead by medical men have revived, suggest the thought how much more numerous must have been such cases in days when medical science knew little or nothing of the symptoms of real death. When it is now ascertained that five per thousand on an average are restored to life who have been certified dead or carried to burial, how many more such cases must have occurred in those times when true physicians and magnetic healers were looked upon almost as gods?

LECTION XXIII. 1-13. -A similar event is recorded in the life of Buddha, where he asks water of a woman, and receives it from a woman of lower caste, who asks how he, of a higher caste, a Brahmin, comes to ask water of one so much lower. It should cast no doubt on this passage.

LECTION XXIV. 1-5. -The cat was an ancient symbol of Deity, on account of its seeing in the dark and otter attributes. More than one instance is given of Iesus' protection of these beautiful animals which in Iudea were, as they are even now in some places, unjustly despised and regarded with disfavour. He, the Friend of all things that suffered, cast his protection round these innocent creatures, teaching men and women to do likewise, and to feel for all the weak and oppressed. This beautiful and much maligned animal was a native of Egypt. But there is no difficulty here, for Egyptian families visited Palestine, and would naturally bring their venerated animals with them, not leave them to neglect or worse, as some "Christians" who ought to know better.

LECTION XXV. 2. -It is remarkable how persistent has been the false rendering of these words in the received Gospels. It is too evident to need any comment. It is not poverty of spirit that Christ commended, but the spiritual effects of literal poverty (not pauperism), which are more frequent than those of abundant riches.

v. 6-7. -Suggestive is this passage of the custom of the Christian Church in building their monasteries and convents generally on high places, the bands of holy men and women therein being truly, in the Dark Ages, the salt of the earth, the light on & hill, without which society would have rotted to the core, and been universally corrupt. The occasional abuses argue nothing against their more blessed influences. Without them our Scriptures would not have been preserved, even in their present condition, and civilization would have been extinct. To the monks of S. Basil and S. Benedict are due the remains of Christianity that have been handed down to us, and by such institutions rationally conducted will Christianity be revived in a higher and purer form, and the Scriptures restored to their original purity, as well as the ancient worship of God. The laxity of some modern monasteries is to be regretted in the matter of flesh-eating, under the plea of health, there being really no such necessity with the abundance of food from the vegetable world as well as animal products. The Carthusian and other monasteries stand as a noble testimony to the healthfulness of the rule when observed in strictness and unabated rigour.

LECTION XXVI. 9. -Meaning; that if the vision be set on one single object and no other, great is the clearness of vision; while, if the eyes be set on number of other objects, the clearness will be diminished with regard to that one.

LECTION XXVII. 2. -The sin of hypocrisy is most loathsome, and the most difficult for those to see who are vitiated by it. To condemn in others the sins we practice ourselves is a common sin of society, which Christ ever reprobates.

LECTION XXVII. 12. -Note the importance which these symbols (including the equilateral triangle) possessed in the eyes of Iesus as illustrations of Eternal truths. The slight mention of these shows the Gospel was written, or addressed to people well acquainted with the mysteries they represent.

LECTION XXVIII. 1-5. -It is easy to see how this would have horrified the mind of Iesus had he lived in these semi- heathen times.

LECTION XXVIII. 15. -To exalt unduly the Christian Sacrament of Baptism, certain words have been interpolated in the A.V. The context shows that such was not the intended meaning, for immediately after, they "justified God by being baptized with Iohn's Baptism."

LECTION XXIX.-The feeding of five thousand with five loaves and seven clusters of grapes has a deep mystical significance, which space forbids to enter on here, but the wise will understand. The two numbers, *e.g,* symbolize Matter and Spirit, Bread and Wine, Substance and Life.

LECTION XXX. 8. -The original Gospels know nothing of the modern doctrine of the Anglican Church. Iesus was "the son of Mary and Joseph, whose parentage we know." This does not contradict, but rather suggests (to reconcile with Church doctrine), the Immaculate Conception of both parents to which the Church is now tending.

Section 4 Lections 31 Thru 40

LECTION XXXI. 4. -Iesus quoted from a more ancient version than we now possess.

LECTION XXXII. 4,5, 8. -The true significance of the bread and the wine in the Holy Eucharist is here taught by anticipation -the substance and life of the Eternal One given and shed for the sustenance of the universe, and this does not exclude, but contains, all other mystical significations which piety suggests, as good, beautiful, and true -each in its place.

LECTION XXXIII. 4.-Here is given the true significance of Ier. vii. 22, or as it should be rendered in that place, " Ye add burnt sacrifice to burnt offering and ye eat flesh. But I spake not to your fathers nor commanded them concerning these things," etc. Else, as translated in the A. V. it is inimical to the sense, see Numbers xi.

LECTION XXXIV. 2.-This beautiful incident does not stand alone in history, a similar story is related of Buddha, the Enlightener of India, the "Light of the East" ; nor is it by any means irreverent to suppose that similar things should happen to persons of similar minds.

LECTION XXXV. 2.-Although these words do not occur as they stand *verbatim* in any version of the Law of Moses as commonly received, the *spirit* of them certainly is there to be found, and in the original copy of the law (the best portion of which has been recovered by spiritual revelation) the very words also. And this original version was doubtless known to this young lawyer, as it evidently was to Iesus, when afterwards he gave the new law to his disciples on the holy Mount when he was transfigured before them in the company of Moses and Elias, the representatives of the old law, which was itself transfigured into the New.

v. 9.-" But one thing is needful "has been interpreted by some, not without reason or probability, as meaning that there were flesh and non-flesh food at the feast, and so he said to Martha, "but one thing (dishes or food) is needful, and Mary hath chosen the better portion." Meaning also, in the spiritual plane, the pure food of heavenly wisdom for the soul. It may have been spoken against luxurious multiplicity of dishes in general. (See Dr. A. Clarke i,1, *loco.*)

LECTION XXXVI. 2-6.-This beautiful story, so characteristic of Iesus, has been most unjustifiably pronounced by modern revisers as an interpolation. It is a parable of human life, ever true, never old. " Let him who is without sin amongst you cast the first stone."

LECTION XXXVII. 8, -That our Lord spoke here primarily of a physical rebirth as the great aid of the spiritual re-birth, there can be no doubt, for he distinctly declares he had been telling Nicodemus of "earthly things" in the preceding words, albeit as the analogies and correspondences of spiritual things, as his usual method was. To interpret this dialogue, even as in the A. V., exclusively of the spiritual re-birth, is contrary to the plain meaning of the words.

LECTION XXXVIII. -"Death" here, as in other cases, is a state of trance or suspended animation, not easily distinguishable from death even by the physician. A circumstance often leading to the revolting fact of burial alive -a fate, however, not so utterly hopeless in the East, where the dead are buried earth to earth in their shrouds, as in the countries of the West, with the modern and barbarous custom of closed coffins, with covers fastened down, and seven feet of earth over them. It is now ascertained by the more advanced and enlightened medical men, and others, and their official reports, that five per 1,000 must, in these English countries, come to this terrible fate, as there are yet no efforts made to prevent it, as in France, Holland, and other countries, where more rational and civilized practices prevail, and where it is found that five per 1,000 come to life, *before* actual interment, or show signs of premature burial after, when exhumed.

The Comments of Section 5
Lections 41 Thru 50

LECTION 41. 11-13. -Very applicable to the present age also.

LECTION 44. 4. -In the Syrian Paschito, accepted by all Christians, in the Aramaic, the very language spoken by the Lord while on earth, the passage reads thus -"Thou art Kepha (rock), and on this rock I will build my Church, and the gates of Sheol shall not prevail against her." In the Aramaic original there is but one word for rock or stone, not two *petros* and *petra,* as in the Greek -on which anti-Catholic controversialists found their views. Rather do these words, "petros" and "petra," being simply the masculine and feminine forms, denote a *duality,* viz. Intuition and Intellects conjoined in the one human Individual; or in philosophy, the inductive and the deductive methods, the one of Plato, the other of Aristotle.

LECTION 45. 7. -The word '" Blasphemy " is derived from the Greek , "blapto" to hinder, retard, obstruct (progress) and "pheme" to speak, thus signifying what is usually known as that conservatism which injures progress and obstructs it. Such a mind is held in bonds by too great deference to authority, or customs, or fashions, or traditions, and closes itself against fresh truths, or the restatement or development of old truths, and thus resists the light which may come through a divine messenger, and commits the "sin against the Holy Spirit" by closing eyes and ears against the higher teaching of God's truth. Such cannot be forgiven *i.e.* obtain release in this age (incarnation} or the succeeding. For the effects of persistent resistance are age enduring, like a prison house of the mind, and it cannot break the strong walls of prejudice and come forth, being fast bound, and needing, it may be ages, for its emancipation from error long cherished.

LECTION 45. 10. -Without slighting any believer's love to the B.V.M., these words clearly shew how Iesus regarded *Righteousness* as relating all true believers to him more than any blood relationship could possibly do.

LECTION 46. 2-7. -The appearance of Moses, to hand over, as it were, the Law and the Dispensation to Iesus, throws a light on the reason of the Transfiguration, which is lost in the accepted version. Elias also appears, so as to make, with Moses, the "two witnesses" required by the law. They recognize Iesus, and witness to him as the great Prophet whom God should raise to succeed and take the place of Moses as the Legislator of the New Dispensation.

The "six glories" may have reference to the six precepts In each table of the law as now given, or more probably to the six attributes of each of the two aspects of the Christ, six feminine and six masculine, as inherent in the Two in One -the "Father-Mother of the age to come." They are referred to in one of the most ancient gospels.

LECTION 46. 10-12 -Compare this with the Law as given by Moses in the "Book of the Going Forth of Israel," p. 35, all chiefly negative precepts. In the law given by Christ there are six negative and six positive. The negative is the external Way, in which certain actions are forbidden. The positive the interior way, in which certain duties are enjoined. As the prohibitions are summed up in the negative form of the Golden Rule, so the commands are summed up in the positive form. There is a very curious recurrence of certain numbers. The words of the two tables being summed up, appear to indicate the feminine, or even number 124, for the 1st Table; and the masculine, or odd number 123, for the 2nd Table; and the numbers 12 and 13 prevail throughout. The words of Moses (see "Book of the Going Forth of Israel ") were 70, denoting an approximate, but a lesser perfection. Had this law been faithfully observed by all Christians, there would have been no divisions or war as now, and this earth would have been a paradise for all. The allegations brought against Christianity and Christ as the cause of strife and bloodshed, are therefore baseless. It is the spirit of selfishness, of perversity, in direct opposition to the Law of Christ, which has been the root of all the evils laid to the charge of Christianity, as he said, "Behold I come to bring peace upon earth, but what if a sword cometh?" Besides the loss of life by wars and intemperance of all kinds, the amount of material wealth wasted yearly in Christian countries, is not only needless, but injurious things, forbidden by the Laws of Christ, is simply appalling. This amount would rebuild the cities of the world, and its ancient Architectural Monuments now In decay, turning them into cities of health and beauty.

Take England alone, the amount of money wasted in flesh meats, etc. (which cannot be had except by cruelty and suffering) alcohol and tobacco (which transform the users into worse than beasts) is on the most moderate calculation (of Mulhall and other eminent staticians) 260 millions. And this increases year by year. (Remember, this was written in 1892 -Ed.) This sum would give employment and food to all who are now without it, and in a little time turn our hideous towns into cities of health, beauty, and convenience, and our waste places into fruitful gardens, and our want and misery into increasing wealth and gladness, It would provide retiring pensions after 50 years of work, for all disabled by age, sickness or accident.

Canon Farrar has shewn that at a low computation, there are 30 millions of destitute now in England. This sum would give a pension of £80 a year to each of such. This sum would also clear the interest on the National Debt (Disgrace) and our Public Expenditure, leaving a surplus of 100 millions.

In this sum is not reckoned the needless, wanton expense of wars (220 millions in two years wasted in S. Africa in pillage and destruction and in plundering a nation of industrious farmers; and now again, in Thibet, the butchery styled by the English "a Peaceful Mission." Nor does It include the money lost in betting and gambling (during the last century 3,000 millions, i.e. an average of 30 millions per year). All these sums are estimated by Mulhall and other eminent staticians whose veracity cannot be doubted, and make a total of at least 300 millions recklessly wasted by this nation alone, all of which would be avoided by simple adherence to the twelve precepts given by Iesus Christ on the Mount.
v. 12. -"Slaughtered" is the truer rendering which has in former editions been wrongly translated "living."

LECTION 46. 24..-The law as given through Moses "by the ministry of angels," and as the Church of Israel (the typical nation) received it through their Interpreters and Scribes, had no precept forbidding cruelty, oppression, flesh eating, drunkenness, the worship of mammon, impure marriages for money or position, and other vices and crimes which are the cause of nearly all the misery which now afflicts men and women "called to be happy sons and daughters of God Almighty, and which are visited by the holy Law on the children to the third and fourth generation of them that despise it, shewing mercy to all who love it." Nor yet did it include that mutual consideration for each other (positively and negatively) which is summed up in the one commandment of Christ "Love ye one another," not excluding therefrom the creatures which God hath given to be our earth mates and companions.

But now by the same Spirit is given through the ministry of angels, a more perfect Law -the Law of Christ- which they knew not, who sacrificed innocent animals (which thing "God commanded not," as said Jeremiah, the prophet, vii., 22), in place of the animal passions and lusts within their carnal minds. And to this end is this Law now given anew after a more perfect way that the multitudes in the Outer Court of the carnal mind may have it written in their hearts, and enter through the Middle Courts of the Soul into the inmost of the divine Spirit where no tables of the Law are seen, for there the Law is WITHIN.

The moral obligations of Mankind have been revealed in two ways; the first and earliest is the negative and external way in which certain actions are forbidden; the second and later is the positive and interior way in which certain duties are commanded. As the prohibitions are summed up in the negative form of the

Golden Rule, so the commands are summed up in the positive form of the Golden Rule, and both in the one command of Christ; "Love ye one another." When these words were committed to writing the monition was given, " Number the words of this Law, for they are the numbers of the Holy City, and of the Solar and the Lunar Year." And so it was found by adding the numbers, and then the figures composing the sums, in four different ways after the manner of the Jewish Kabbalists, shewing that the same mode of reckoning in their English Version was retained by the translators for the purpose of (as far as possible) preserving the very words of the Law in their purity and integrity.

LECTION 47. 2-7.-There is hardly a law that has been given by man or by God, but has been abused and turned into an instrument of oppression by the perversity of man, evading the spirit and insisting on the letter alone.

vv. 10-16.-This parable of Dives and Lazarus is pregnant with deepest teaching, utterly setting aside the narrow dogmas of the sects which have been founded on the traditional interpretation of it in the A. V.

LECTION 48. l-3. -The feeding of one thousand with five melons is another instance of the love of Iesus to the hungry and thirsty thousand who came to his ministry. He first feeds the body, then instructs the soul. Too often it is the reverse of this among his followers. Hence, "The lack of love causeth many to wax cold."

LECTION 49. 7. -Iesus here disclaims the sacrifices of bleeding victims as being by part of true religion, but rather contrary thereto, adding to the guilt of the offerers. How often religion, so called, is put before morality, and dogma before loving-kindness and mercy.

v. 8. -This is fulfilled in the Catholic Church of East and West, in the arrangement of the Sanctuary and Choir, in which, however, the standing altar of incense has been replaced by swinging censers -the one should not exclude the other. Each has its place and use.

The Comments of Section 6
Lections 51 Thru 60

LECTION 51. 2. -Many people think that perfect freedom is the power to do wrong as well as right. Such is not Christ's teaching. Freedom is the power to do moral good, nothing else : the other is not freedom, but slavery to the evil nature. This is the teaching of Rosmini and of the Franciscans, and is evidently the teaching of Christ. Other animals than man have the freedom essential to their nature, which, if they are allowed to follow, is a kind of moral good. In a true state of nature, the other animals are found innocent, till corrupted by the cruelty of man. Herbert Vivian (" Land Of the Lion of Judah ") writes :-" On the beaten track I was much impressed by the fearlessness of nearly all the animals I saw. The first time I tried to stalk a herd of antelopes I gave myself a great deal of unnecessary trouble, dissembling behind bushes and reserving my fire lest the first shot should irretrievably disperse my quarry. I found, however, that as a rule directly they became aware of my presence they turned round to look at me and would remain while eight or ten shots whizzed about their ears, not always bolting, even when one of their number had been laid low. Nor did they ever scuttle away very far. They would disappear over the ridge of a hill and wait within rifle range of its summit as if for me to try my luck again. The smaller animals would be more fearless still, and might often have been knocked down with a stone or a stick. There were numbers of pretty little grey and white squirrels with long bushy tails; they would run to pick up a bit of bread when I threw it, and lit up a few yards away from me nibbling it with both hands. One day I thought of shooting a jackal, which was hovering about near my camp, for I had heard that the skin of a Somali jackal is worth having. I took up my gun and strolled out to get an easy shot. My servants, thinking to help me, whistled to it, whereupon it turned round and looked at us as though to see what we wanted. Then *I put away my gun, for I had not the heart to shoot.* It would have been like killing a house-dog." *(See* also "Darwin's Voyage on H.M.B. *Beagle,"* where a similar testimony is given). If man deprives his essential knowlege of good, he disarranges in himself all that part of creation that is within him, and becomes lower than the beasts of prey.

The Gospel Of The Holy 12

LECTION 52. 4. -The testimony of those who saw and knew Jesus as to his age, has been strangely ignored by writers of Biblical history and by the Church in general. This matter is briefly discussed elsewhere in these Notes, and deserves the attention of every student and thoughtful person. (See Notes liv.14-16; xcv.9.)

LECTION 53. 3. -The healing of the blind by means of clay mingled with saliva is mentioned by ancient physicians. Vespasian is said to have cured by this means. This shows that Jesus did not hesitate to employ natural remedies, when they were likely to effect their purpose.

LECTION 54. 1-13. -The wrangling of the Pharisees over this case of healing has its parallels in our times in the Churches which assign to the devil all that they cannot comprehend, and cut out the Healer as a sinner and a heretic, denying the power of God in Man.

LECTION 54. 14. -This is one of those "parables and dark sayings" of him who spake as never man spake. The words taken literally suggest to the mind a perfect crystal sphere, and by correspondence, a perfect man or woman- in modern phrase "an all rounder," one who views things not from one side only, but from every side. There are many who keep the law in one or more points, but neglect all the rest; or keep it in all points but the one which is against their own particular failing -who "compound for sins they are inclined to, by damming those they have no mind to." But few are they who teach, and still fewer who practice an all round obedience to the laws of Christ. Many are they who loudly condemn one or more forms of evil, in order that they may more fully indulge in some other, in which is their corrupt taste; or condemn little errors in others that they may escape notice of their own greater breaches of the law of loving kindness to all creatures. *(See* also Lection lxviii.,18, and lxix., 5.) On which otherwise obscure passages the above remarks may throw a needed light. In these the conditions of the individual are referred to, the one throwing a light on the other. For as with the Church so with each individual composing it, they must progress to this perfection of character -the "measure of the stature of the fulness of Christ." The letters beneath were found on an ancient English stone, cube-shaped, "six squared," in a church in Warwick, with these lines, "I, Thou, He, She, We, Ye, It, They ; All are one in Me "-supposed to be the work of a Rosicrucian Pantheist of old time.

LECTION 54. 17-20. -The meaning of these words and this action is very obscure, but if we describe the magic square of 7, it seems to make it intelligible as the mystic symbol of him who regarded everything by number and by measure, and which seems to have reference to the period of his mortal life, 49 years, as well as the number of the Council, Cardinals and Priests of the Church universal, 48, presided over by its Head, 49, which the action of Iesus seemed to symbolize, and in a way, foreshadow.

LECTION 54. 17-20. -Here we have the original words of Christ, from which Paul adopted his simile in Rom. xii., and In 1 Cor. xii.

MAGIC SQUARE OF SEVEN
Sigillum XTI

22	47	16	41	10	35	4
5	23	48	17	42	11	29
30	6	24	49	18	36	12
13	31	7	25	43	19	37
38	14	32	1	26	44	20
21	39	8	33	2	27	45
46	15	40	9	34	3	28

21 Summations of 175. Sum total of terms, 225.

This most ancient seal of XT - the High Priest, the At-one-er and At-one-ment of all things, the Lord our Lady Jesu-Maria - is one of the most ancient and sacred symbols of the XN religion, extending back to pre-XN antiquity, and symbolizing the restoration of the primeval order to the universe, bringing order out of confusion. In this most curious and mystical combination of numbers we have 21 summations of 175 (this number cabalistically added being symbolic of XT amid the XII.), viz., 7 rows; 7 columns; 2 diagonals; 1 central cross, which sums 2 ways (5 vertical with 2 transverse numbers, or 5 transverse with 2 vertical) ;3 squares made of the 4 corner numbers of the inner, middle and outer courts with 3 vertical, or 3 transverse central numbers. There are other curious combinations, *e.g.,* the sum of the four corner figures of each of the squares being added give 100. Likewise the extreme end numbers of the central cross, while the complementary pairs, wherever found, being added, sum 50. Regarding the symbolism, each number denotes some place or office in the council of the Church, whose symbol is 120, under her head or his vicar. 49 symbolizes the number of the priesthood of the Universal Church under his head, and 25 that of the Local Church under his head, while 1 represents the Unity underlying all. It will be noted that 49 symbolizes also the number of years in life XT by the testimony of His contemporaries, the Jews, and S. Irenaeus, B.M., A.D. 189, who received it from the surviving apostles and disciples of the Lord. His ministry having lasted 18 years from the date of His baptism, when He was manifested as the XT, the beloved son of God.

I.H.W.Y.S.T.

LECTION 55. 1.-This beautiful parable bas been sadly mangled in the A.V., and shorn of the opening incident which led to the discourse.

LECTION 56. -This touching account of the raising of Lazarus is here given as it took place. The verses 13-16 in the Authorized Version are an evident interpolation to magnify the occasion, for, being omitted, the narrative is unbroken and complete without them. As with the daughter of Lazarus, so with Lazarus, he was carried to his burial in a state of trance, indistinguishable from death, and by his friends believed to be dead. At the present time in countries where there are mortuaries or waiting rooms for the dead, it is found that five per thousand recover on their way to burial who otherwise would have been buried alive.

LECTION 57. 4. -The doctrine of guardian angels receives full support from these words. But the Churches of the so-called Reformation have flung away this consoling and helpful belief, with other doctrines of the Christian Church in all ages, the truth of which science and occultism are now showing.

LECTION 58. 2. -The charity and comprehensiveness of the true doctrine of Jesus here manifests themselves. It is not a mere narrow creed or belief, but true repentance which merits the forgiveness of God.

LECTION 59. 1l-12. -The teaching of our Lord as to cycles, and the unity of life, in many existences, has been suppressed for long ages, but now sees the light, at the end of the cycle.

LECTION 60. 16. -The same Zaccharias who is mentioned in the beginning as the father of John the Baptist (see Note 111-2), also the Proto Evangelism attributed to James, the Bishop of Jerusalem.

The Comments of Section 7
Lections 61 Thru 70

LECTION 61. -All through this chapter the language is highly symbolical, but will present little difficulty to the initiated. *v.* 12. (*See* Note in the original "Genesis" Edited by same Editor).

LECTION 62. 1. -This parable of the ten virgins most accurately indicates the oblivion and indifference which shall come on Christians in the last days of the Christian Church -the days of Laodicean indifference, the Seventh or last age.

LECTION 63. 8. -These words are one of the "last sayings" of Jesus and vividly describe the duty of a Christian Council, so oft neglected.

LECTION 64. -The occult teaching, In this discourse, of Jesus to his twelve has been handed down in spirit through the ages, but the world is blind and perceives not. See the same teaching in "New Light on Old Truths," founded on this Scripture.

LECTION 64. 8. -Beneath this profound saying of the Ghost Physician, the student cannot fail to notice the intimate and correct knowledge of the human frame, underlying the spiritual truth, which he enunciated. This knowledge has been claimed by science only some centuries later. The inner self- "alternate sex," in every man and woman, which occasionally manifests itself in the dream state, seems to be no mystery to him. (*See* G. Leland on " The Alternative Sex"; Welby, London).

LECTION 65. 2. -It has been supposed by some, and not without some reason from the words of the Gospel, that envy and jealousy, and not greed of money, were the cause of Iudas' treachery, because he desired Mary Magdalene, and she had given all her love and devotion to her Master. This inner feeling seems to be concealed beneath the cloak of zeal for the poor. "From that hour he sought to betray him." It is as probable that all three motives urged him, as they do the multitudes nowadays, who grudge magnificence of architecture, music, etc., under the cloak of "unity." "These things ought ye to have done and not left the other undone." But such show their hypocrisy by their reckless contributions to war, and to all manner of pleasures and amusements and luxuries which minister to self. By the spirit of Iudas Iscariot are all such led and dominated.

LECTION 68. -Again, studying these "dark sayings" so difficult to understand, recourse has been had to certain figures known to the early Gnostics. *(See* the "Squares and Circles" by the Editor of this Gospel.) The Magic Square of 11 has been found wonderfully to explicate, symbolically at least, to the mystical, the meaning of the passage. The form exactly illustrates what the Lord in symbol taught to his disciples of the bringing forth of order out of disorder, perfection out of imperfection, and out of deficiency fulness. Compare the Magic Square given below with the natural square which anyone can form by writing the numbers in consecutive order; the result is at once seen, and may help to arrive at the meaning of this very mystical passage:-

MAGIC SQUARE OF ELEVEN PERFECTED
(Esslesice Militantis Sigillum.)

LECTION LXVIII.—Again, studying these "dark sayings" so difficult to understand, recourse has been had to certain figures known to the early gnostics. (See the "Squares and Circles" by the Editor of this Gospel.) The Magic Square of 11 has been found wonderfully to explicate, symbolically at least, to the mystical, the meaning of the passage. The form exactly illustrates what the Lord in symbol taught to his disciples of the bringing forth of order out of disorder, perfection out of imperfection, and out of deficiency fulness. Compare the Magic Square given below with the natural square which any one can form by writing the numbers in consecutive order; the result is at once seen, and may help to arrive at the meaning of this very mystical passage:—

MAGIC SQUARE OF ELEVEN PERFECTED.
(Ecclesiæ Militantis Sigillum.)

56	117	46	107	36	97	26	87	16	77	6
7	57	118	47	108	37	98	27	88	17	67
68	8	58	119	48	109	38	99	28	78	18
19	69	9	59	120	49	110	39	89	29	79
80	20	70	10	60	121	50	100	40	90	30
31	81	21	71	11	61	111	51	101	41	91
92	32	82	22	72	1	62	112	52	102	42
43	93	33	83	12	73	2	63	113	53	103
104	44	94	23	64	13	74	3	64	114	54
55	105	34	95	24	85	14	75	4	65	115
116	45	106	35	96	25	86	15	76	5	66

33 Summations of 671. Sum total of terms, 7,381.

LECTION 69. 8, 9. -This saying of Iesus is very difficult to the popular mind, as apparently reversing the original injunction in Gen.1-3, "Be ye fruitful and multiply." To understand this, it must be borne in mind that the promise of a Messiah to redeem the world, has, from the earliest times, begotten in the woman of the Hebrew nation, to whom it was specially given, that insatiable desire for offspring, each woman thinking of herself as the possible mother of Him who was to come and save. Iesus, the true Prophet, seeing the tendency to the propagation of the unfit (as we now see) to bring want, misery, squalor, vice and crime, through the inability of most parents to bring them up as they should be, owing to the curse of competition and greed, here proclaimed to Salome, in answer to her query, that he would reverse all this tendency among his followers, and through them, extend this reversal to mankind at large, for He, the desire and hope of nations, having come, there was no longer any supposed necessity or reason for such increase, of which He, the Prophet of God, fully foresaw the evil, in the ages to come, as we now fully experience it.

LECTION 70. 1-5. -Long has Iesus suffered reproach, for those words so falsely attributed to him, in place of the impulsive Peter, who spoke them, and with whose character they were in full harmony.

The Comments of Section 8
Lections 71 Thru 80

LECTION 71. 1-4. -Twice the Lord is said to have performed this symbolic act. Surely, at his return, it will be his first work! For since the first ages till now the spirit of the world ruleth, and mammon is dominant, and every kind of wickedness in the name of religion, zeal for purity, etc.

LECTION 72. 1. -In the language of the Churches of this day, there is but one mansion in the Father's house, and *that* is claimed by each of over 300 different sects as its own, and all outside are damned, not for their evil deeds, but because they cannot see as their rulers profess to see.

LECTION 73. 1-6. -"I the true Vine, ye are the branches" -in unity with the stem by the continual possession of the One Life, not by *mere* external unity, valuable as this is, and certainly not by a dead uniformity of opinion in all things. *"Tot homines tut sententice."*

LECTION 75. 1. -Jacob is the same as Iames -called "the great." Nathanael is Bartholomew. There is no proof that Jude was the same with Thaddeus, as is alleged by some. The number at first seems to have been twelve exclusive, or thirteen (to the world's eye) including Iudas Iscariot, till he should manifest his falsity by his treachery, when he went out directly before the holy supper, leaving Iesus with the twelve -the complete number of Apostleship, which, being even, admitted of no one among them being "Master," save Iesus, who was over them.

v. **2.** -Whether the appearance of the Master and his disciples in symbolic festal garb may not have been seen only by the spiritual eye of some of the disciples or not, the lesson is the same. Reverence and love of beauty and order are to be seen in God's House -symbols of the glorious garments of that Being Who is the Eternal Mystery and Beauty manifest in all things.

vv. **15, 16.** -That Twelve is the complete number of the Apostleship and that Iesus sat down *"with his twelve"* at the holy supper before his crucifixion, seems evident from the received gospels, and still more so, from the fragments lately brought to light. Iudas Iscariot appears then to have been *among* the twelve but not of them, therefore before the Eucharistic rite is celebrated "he goes out." If there were any ill omen at all about the number thirteen it would therefore be thirteen as the number of Apostles present, exclusive of the Master and Head. But to thirteen, inclusive of the presiding host, no ill omen could attach, but the reverse.

LECTION 76. 4. -There are two other alternative versions of these circumstances of the last supper in the A. V .-First, that of St. John who, in the received version, expressly affirms that Iesus was crucified on the very day of the Passover and consequently the Eucharist was instituted the day before and not on the feast day Itself' and the Passover was on the morrow after the trial on the day of the crucifixion. Secondly, that of the hree other gospels, which all affirm that the Eucharist was Instituted on the Passover the pascal lamb was slain. If the latter, it must be remembered that the Essenes (of whom Iesus was apparently one), were by Jewish regulation allowed a separate table at which no lamb or other flesh-meat was eaten, as they were vowed abstainers from blood sacrifices and the eating of flesh. If the former it was not the Passover at all, and Iesus was not bound as a Jew to eat of a lamb. In neither of these cases, therefore, was Jesus under the alleged necessity of killing a lamb and eating of flesh-meat in order to fulfill the law. In any case the causing of an innocent lamb to be killed and the eating of such is contrary to all that is known of the character of Iesus the Christ, whose tender love extends to all creatures. If Iesus was not an Essene, then nothing can be said against the accuracy of this version of the holy supper, and the charges brought against him in the account of the trial as now given by the Spirit.

v. 9. -"Bread," *i.e.* unleavened cakes of pure meal such as in use at the Passover. "Wine," here and through the Gospels, as used by Jesus and His disciples, means "the fruit of the Vine." which is pure wine mingled with four or two parts of pure water, the latter mystically representing the humanity, and the former the Divine Spirit. The strong fermented wine of modem use was never used on such festive occasions, nor even generally, except thus mingled with water. It is to be noted that the Saviour consecrated the Eucharist by Invocation of the Holy Spirit, and this has been faithfully followed by all Churches of the East, the words of institution being merely recited before, as a historical preamble, giving the authority for the action, and in no case as the words of consecration, according to the corrupt use of the West.

v. 13. -In the received Gospel Iudas is consigned to eternal perdition, but it appears rather that he who was all compassion and prayed for his murderers, prayed also for the man who was overmastered by his passions, blinded by envy, jealousy, greed of money, or, as some say, by desire to push matters to their conclusion, and procure some decisive miracle that would establish the claim of his Master to set up a temporal kingdom.

v. 26. -It is not stated whether there was any musical accompaniment, as is usual in the religious dances and processions of the East, but if so it was probably of the simplest, such as the Pipe, used on such occasions.

v. 27. -The Mazza, or unleavened cake, to which may the word "Mass" be traced as applied to the Eucharist, or "Breaking of Bread" -but preferable perhaps is the interpretation of "ite missa est " -the oblation (= prayer) is gone, *"sent up."*

v. 30. -Here was perhaps more probably the sole motive actuating Judas -his ambition- the desire to see a miracle, and the early sovereignty set up before the time.

LECTION 77. 2. - Here the Lord addresses Simon, not Peter. In the A. V. confusion has arisen owing to the same name being given to two Apostles, and Peter is made to reply. It does not seem likely that one who thrice betrayed the Lord should by him have been placed in the highest authority, as it subsequently appears that Peter was.

LECTION 78. 12-18. -The belief that Peter denied his Master is probably owing to two of the Apostles bearing the same name, Simon Peter and Simon the Canaanite. Here we are given the right version. The error is one that might have been easily made. It is worthy of notice, that this ancient Gospel attributes to Simon (not to Simon Peter) the thrice denial of Iesus, and his fully exonerates Peter from the baseness generally attributed to him, and to which there is no allusion in his writings, but rather the reverse in the accepted gospel, where he was first to draw the sword in defence of his Master.

LECTION 79. 2. -In a preceding Lection (LII.) the Jews at that time adjudged him then to be forty-five, and here Caiaphas, who must certainly have known his age, declared him to be " not yet 50," *ie.* about 49. This is borne out by the A. V. and by the testimony of S. Irenaeus, A.D. 120-22, and the testimony of S. Iohn the Apostle and his immediate disciples.

LECTION 80. 1. -The heading of this Lection in the A. V. is most misleading. "Penance," implying reparation of some kind (even though not of the right kind), is the more correct description of the act.

The Comments of Section 9
Lections 81 Thru 89

LECTION 81. 9. -These words, or the substance of them, are also to be found in one of the gnostic Gospels, which record many genuine sayings of the Master.

LECTION 81. 2. -This verse, suppressed by corruption of the Gospel, doubtless refers to the keeping the Passover within the gates without the slaying of So lamb, a capital offence by the law *(See " New Aspects of Religion," by Dr. H. Pratt)*, or it might refer to keeping Passover the day before. There is much uncertainty on this point, the Gospels in the A. V. setting forth two different views, mutually contradicting each other, but neither of them implying necessarily the eating of a lamb by Iesus and his Apostles.

LECTION 82.10-12. -Eli Reclus, a French writer, has some interesting remarks on the rite of human sacrifice as practised among the Khonds from time immemorial. The coincidences in the details are very striking, shewing the similarity of superstitious ideas in all countries and tribes of the primitive world-ideas which survive even in our own times " civilized ., as we boast them to be, but in reality savages when the skin deep "civilization" and culture are suddenly brushed away by some violent popular outburst, a in England, against the peaceful peoples of the Transvaal by which she brought herself to the lowest depths of infamy, and unwittingly clothed herself in the colour symbolic of dirt and mire.

v. 20. -In the Gospel attributed to Peter there is mention of the same circumstance. And to bring to mind, by symbolical art, this awful scene, among other reasons, the dark unbleached candles are lighted on the Altar on the day and at the hour when the Church commemorates the crucifixion of the Redeemer by an ingrate priesthood and people, when the light of the sun is shut out or obscured, and the chancels are draped in black.

v. 30. -It should be observed that in this Gospel, the mystically central organ of the Sacred Body, the ..Heart " is emphasized rather than ..his side," as in the A. V. on which last reading the strange custom of having a side entrance or porch to Churches is alleged to have been founded. The traditonal but corrupt reading of Gen. vi.16 has doubtless originated the error. *(See "Original Genesis.")*

v. 31. -They pierced his Sacred Heart with a spear, and this is symbolised in Christian Churches (which are generally cruciform either externally or internally where they are not circular), where the choir *(Cor.)* is in the intersection of nave and transept, and the altar of incense is *(ought to be)* in the midst under the great dome, symbolizing that the Sacred Heart of the Crucified is venerated from the centre to the extreme limits of Christendom-the Heart of God which embraces all creatures in its boundless love.

LECTION 83. 5. -It has been maintained by some with no small degree of reason and probability, that the day of Crucifixion was not Friday, the day now observed by Christendom, but Wednesday (mid-week), by which date alone would be truly fulfilled the prophecy of Daniel, and the only sign of the truth of His mission which he would give to his generation. There shall no sign be given it, but the sign of the prophet Ionas, for as Ionas was three days and three nights in the whale's belly, so shall the Son of man be three days and three nights In the heart of the earth." Against this plain testimony there is of course the canonical record as we now have it including the frequent explanatory notes which may have been incorporated in very early times from the margin into the text, or interpolated in ignorance of the original script, which no man living has ever Been from this to the 10th century when all manuscripts were in the hands of the religious orders of the Church, and from them proceeded. If these words of Jesus be a genuine portion of the Gospel, as all admit they are, those notes of time, in the present accepted Gospels must be spurious, or the work of scribes who sought with honest and pious intent to harmonise the words of Scripture with the existent beliefs and observances of their age. In the gospel as now given there is absolutely nothing to militate against either of these views except the words of Jesus above cited, which cast the weight in favour of this chronological arrangement which interferes with nothing of Christian doctrine. Sunday, as now the day of his public entry into Ierusalem, preceded by the last anointing by Mary Magdalene on the eve before it. Monday, the day of evil counsel. Tuesday, the day of the Pascal feast of Christ. Wednesday, the day of the crucifixion, if not of the actual Jewish Passover. Thursday, Friday, Sabbath days of watch, of mourning and vigil. Sunday the day of the Resurrection, midnight or 8 a.m., early dawn " (after three days and three nights were fulfilled) and of the rising of many who slept and of their appearance in the holy City."

LECTION 84. 6. -0n this passage the celebrated writer M. Renan, bases his assertion that "but for Mary Magdalene Christianity would never have existed." It was she who first proclaimed the central fact -the Resurrection of the Lord. There is a true and a false side to his words.

LECTION 85. 15. -These words though not fully given in the A. V. have been made the basis of an attempted explanation by M. Renan, who could not receive the alternative view that the body of Jesus was dematerialized, rose, and then appeared in spiritual form, which view is held by believers of modern manifestations.

LECTION 86. 4. -"The sacrifices ceased that day" -here is not meant for any permanence (as generally believed) for they went on, we are told, for forty years, till the destruction of the Temple by the Romans.

LECTION 87. 1. -The power to come in, or to go out through closed doors, has been shown in modern times to be no impossibility, but a proven fact in psychological phenomena. The words here do not *necessarily* imply that such manifestation took place. It is not said, "they were locked," but the power of the Spirit to materialize, and dematerialize, Bond appear in human form (under certain conditions) is too well known to be denied." Report of Dialectical Society on Spiritual Phenomena," etc.

LECTION 87. 2. -The contradiction in the A.V. is here no longer seen. That a spirit can *appear* in flesh and bones has been testified over by thousands of competent witnesses in this as well as other ages. There is no death, and the returning spirit can *appear* in any form. Of these things we are witnesses.

v. 8.-A similar passage to this occurs in the "Pistis Sophia," an ancient gnostic Gospel.

LECTION 88. 5. -Most affecting is this, the experience of all who in this world of madness and unreason attempt to declare the whole counsel of God. It broke the heart of Jesus, it crushes the heart of every prophet or apostle worker for good, filled with his spirit- " Jerusalem, Jerusalem I would - but *ye would not."*

LECTION 89. 2. -" That night they caught nothing"-henceforth their labours were to be in the Spiritual Kingdom to save souls -not destroy them- by bringing them within the Church of Christianity, from barbarism and darkness to reason and light and love.

The Comments of Section 10
Lections 90 Thru 96

LECTION 90. 4. -The art of cutting and polishing glass and stone was well known in Phoenicia and Egypt, before the Christian era, and in Pompeii numbers of such crystals were found in great variety. It is a beautiful symbol appealing to the mind.

LECTION 90. 12. -Our Lord never damned or blamed those who could not see the divine truths, which he taught, and receive them. He had patience with them, as being without the fold, without light, and not admissible to the Kingdom, so long as they remained in their darkness and impenitence and self-doomed to eternal death if they persisted.

LECTION 91. 5. -The idea of baptizing unconscious infants seems never to have entered the mind of Jesus. He blessed them, but he also blessed other animals, and things that had no sentient life. Baptism implies belief and confession of faith and repentance from evil works and ways.

LECTION 91. 6. -0ver 2,000 years before Christ there existed on the shores of Lake Meeris, in Egypt, a labyrinth of seven circular wall-enclosed winding paths, represented by Boticelli in one of his engravings, which we here reproduce adapted for Christian rites. This was used by the Egyptians in their initiations as a symbol of life, and the wanderings of the soul in the flesh, till "seven times seven" times purified and meet to appear before God. There appears to have been a similar one in Ierusalem before the demolition ordered by Hadrian, and this may well have been used by the Early Christians in receiving candidates for admission into the Christian Church. In after ages, this idea seems confirmed by the remains of Labyrinths to be found at the west end of several Churches in Europe. A beautiful specimen at Chartres of dark stone, inlaid with light, the winding path of about 666 ft. round to the centre shrine. They were in later ages used as places of pilgrimage or of penitential exercises during Lent and other seasons, but there is no question that the original intention was to symbolize to the penitents *the manifold wanderings of the soul in the outer* darkness before, being purified, it reached the beatified abode, the marvellous light of the Divine glory , indicated by the central Shrine, whose pavement and walls were of golden colour, and illuminated by many lights. In this shrine was situated the font, descended by seven steps, and the altar at which the candidates were received after their baptism. These sentences, supposed to be part of the rite of initiation, shew this to be the case: "Going out from My presence ye shall wander in the outer darkness, but in due time ye shall return, and seeking Me through repentance ye shall find Me, who am the Light of all who seek." Again "A Pilgrim am I, wandering from my God through the darkness of the world, I desire to return to my ancient home, whence I came, to see my God who is my light and my joy." Again, "Forty years and nine, yea seven times seven doth Israel wander in the wilderness of this world, for it is a generation that do err in their hearts, knowing my Holy law, and not obeying. But those who shall obey my law and overcome the evil with them, are made perfect. They are made pillars in the Temple of God and shall go out no more;" and again, "Glory be to Thee O God, who bringeth us out of darkness into thy marvellous light." All along the walls run sentences from the 78th Psalm describing the wanderings of Israel, which, entering in with lighted taper, the catechumen is supposed to recite to himself.

This building may comprise part of the church attached to the western end, or separate from it. There remains no trace of it at the present day, nor is any description, like many other things that have been forgotten in the darkness of the early ages of persecution and desolation, like the ruthless tide of destruction which prevailed in our own country in the fifteenth and sixteenth centuries and from which it has never recovered.

LECTION 91. 7-8. -In the Editor's former work "Palingenesia, or Earth's New Birth," 1884, incorporating some Ideas from this Gospel (part of which he had then received) these two rites referred to, by some oversight were transposed. Here, as in" Church of the Future" 1896, by the same Editor, the correct order is given. It is at present out of print.

LECTION 92. 4. -Here we have further proof, if any were needed, that Jesus was brought up in the tenets and customs of the Essenes. See "Christianity and Buddhism" (a remarkable book by Arthur Lillie) for the full discussion of the subject.

v. **6.** -Similar were the rites of Mithra. From the days of Noah and Melchizedek these pure mysteries were celebrated -though not in the fulness of the light of Christ.

LECTION 94. 7. -From this, as from other words of the Master on previous occasions, it is evident that his servant Paul borrowed from him many of the ideas, and similes and wise sayings scattered through his Epistles, and not Paul only, but also the other Apostles. *(See* also verse 9).

v. **10.** -It has been alleged that the laying down of rites and ordinances for Christianity has been the cause of division and strife in all countries. Nay, rather have not these divisions and dissensions been caused by the omission of the directions given by the One Head acknowledged by all during the period between his resurrection and ascension and the generation immediately after, and the handling of them down by that tradition so liable to corruption in place of the written record. But much more were these divisions and dissensions caused by the interpolation of dogmas not making for goodness and unity, by the suppression from the records of the vital essence in the holy law given by Iesus on the Mount, which, had it been preached and known and obeyed by all, would have made the earth a paradise in place of a hen for the weak and the helpless.

LECTION 95. 5. -There is no doubt that the "power" here referred to means the spiritual power which we read of as exercised by the followers of Jesus and other great prophets in all ages more or less. Taking the various accounts in the Gospel and ecclesiastical history as correct, miracles *(i.e.,* wondrous works wrought by the exercise of faith and will power and often by the uses of subtle forces of nature, quite natural, but seemingly supernatural to those in ignorance of these forces) were of frequent occurrences in those days, even as they are in these days, but better understood, false miracles being no proof of the non-existence of true ones. Often they would be the effect exercised on the minds and imaginations of vast numbers of the poor and afflicted, the diseased and suffering of humanity by faith in some great champions of the oppressed, themselves destroyed by the oppressor, yet realised by faith, if not by actual knowledge as still living and acting, with hands outstretched to heal and bless those who invoked their aid.

v. **9.** -From the testimony of the Jews, John viii. 57, A. *V.,* it appears that Jesus at that time was not far from fifty years of age, and this is supported by S. Irenmus, 120-200 A.D., who appeals to the gospel as received by those of his day and to all the elders as testifying the same," those who were conversant in Asia with John, the disciple of the Lord, affirming that John conveyed to them this tradition." "Some of them," he says again, "not only saw John but the other Apostles also, and heard the very same tradition from them. Bond bear testimony to the truth of the statement."

The Editor of this Gospel has been credibly informed by an esteemed friend of his, "a Syrian Bishop," and a relative of the late learned Pope Pius IX., that he frequently (in private) assured him that he firmly held this (as a private opinion), the present time (1870) not being yet ripe for a public declaration on this and similar subjects, now introduced into the notes to this and other publications of the O.A.

LECTION 95. 8. -Mary Magdalene was chosen by our Lord as a type of the Church, in her fallen condition, redeemed by His love, and would be fitly one of the first fruits taken to be with her Lord, as Ioseph and Mary were after. She was the constant companion of Iesus' Ministry, to him she ministered of her substance, she anointed him for his Ministry, and for his Burial. She was the last at the Cross, and the first at the Tomb, and to her aJone He gave the commission, " Go tell Peter," and wheresoever the Gospel was to be preached, her love and devotion to her Master were to be declared.

LECTION 96. 1. -This number, 120, has many mystic significances, and was foreshadowed by the number of souls saved in the Ark at the Flood ("The Original Genesis"), which included 48 *(i.e.,* double 7 + 34) + 72, a number of deep mystic significance.

v. **2.** -The manifestations described here have been repeated in modern times. What God does in one age, whether by angels, spirits, or adepts in the flesh, the same unchanging God repeats in another. Whether the miracle respecting the preaching of Peter took place in the persons of the Apostles, or in their hearers, we have no means of ascertaining, but the fact remains. Most probably in the hearing of the hearers, so that each was enabled spiritually to understand. or else all were moved to speak and to hear in a tongue com

LECTION 96. 9. -These words would seem to suggest 50 years at the natural term of the working life of men and women, after which they should be relieved from the necessity of further labour. Having worked for six weeks of years, man is entitled to the Sabbath of Rest from further servile labour.

v. **5.** -That these material symbols mentioned were actually used in this assembly may be doubted by some. It is more likely that the disciple, who had the gift of spiritual sight and hearing developed, heard and saw some of the things here described, as symbols of deep inner truths. The expression "to none was given aught that was not given to all," seems to shew this to be the case, and that It was understood in a mystic and spiritual sense.

LECTION 96. 12. -From the earliest times the Ecclesiastioal grades and orders of the Christian Church were, in the Local Church, seven in number. 1, The Angel or President. 2, The Presbyter. 3, the Deacons. To these were added 4, Readers. 5, Singers. 6, Acolytes. 7, Door- keepers. Each of these, from the Angel or Bishop downwards, had his help or coadjuter to represent him in his absence. And among the Laity were, 1, The Faithful. 2, the Penitents. 3, the Catechumens. The Angel or President of the Local Church was a Pastor in the Universal Church, in which the Supreme head, under Iesus Christ, his Vice. regent was the Angel or Bishop of the Universal Church and under him 12 Apostles, 12 Prophets, 12 Evangelists, and 19 Pastors, constituting the higher Priesthood of the Universal Church. Each with his coadjutor, as in the Local Church, to represent him in his absence. Under these were the Deacons, one to each Priest, 48 in number. These constituted the supreme Council of one hundred (the supreme Angel or Pontiff having two additional Coadjutcrs chosen by him). But there is a higher meaning, mystical and doctrinal.

LECTION 96. l3. -Here we have suggested the earthly Trinity of Jesus, Mary, Joseph; the reflection on earth of the celestial Trinity of Father, Mother, Son-daughter ; and so clearly as to avoid all idea of worship, *latreia,* to created beings. Nothing in the text is affirmed or denied, but simply "they were not," "the Lord took them" even as God is said to have taken Enoch, and as he took Mary Magdalene, Jesus being first of first-fruits and Mary mystically one with him, as Christ and his Church are one. Then Mary and Ioseph followed after (whether in flesh or Spirit we are not told) to be with him, whom they venerated here on earth. If the assumption of Mary his Mother has been defined by the Church (to guard the doctrine of the true Divinity of Christ) shall not also the assumption of her Immaculate spouse? The authority in the four accepted Gospels is, for the B. V .M. no greater than it would be for Ioseph in this case. Already the Church sings of Iesus, Mary Bond Ioseph as the reflection, the shadow on earth, of the heavenly and eternal Trinity of Father, Mother Bond Holy Child. The immaculate Virgin and her spouse represent the great assembly of the just made perfect-the regenerate compa.ny of the sons and daughters of God Almighty, in whom there is no spot or stain of imperfection, the Bride of the Lamb, of whom the Temple on Mount Zion was an earthly symbol.

LECTION 96. 18. -This is most probably the oldest of the creed forms, if for a "creed" it was intended, long lost but now restored. It clearly expresses the faith and the practice of the Early Christian Church (reaching back, perhaps, in some truths, to the days before Noah), which we see glimpses of elsewhere in the writings that remain to us. The decree of the Church in her first general council at Jerusalem, founded on it as a deep psychological insight, and given under the direct influence of the Holy Spirit, is well known, though generally forgotten or ignored by Christians; for the logical deduction is total abstinence from **flesh meat**, which can only be obtained for food by the destruction of innocent life, whether by bloodshed or by strangling. And such destruction of life for selfish ends is placed with fornication and other deadly sins. Some portions of the Epistles of S. Paul in the A. V., notwithstanding all these, are manifestly interpolations of a later date after the discipline of the Church had been relaxed, and the evil customs Bond fashions of the world had sought and gained admission in the Church in union with a corrupt state. Both this symbol of Faith and the 12 Precepts of Jesus have been published before in a separate leaflet form in the year 1881.

v. 22. -It has been asserted by Church historians of all shades of opinion that incense was not used in the Christian Church till the fourth century. Here we have evidence of the contrary, and proof that in the early Church of Jeruslem It was in use at the time this Gospel was written. This with other ceremonies, ornament., etc., not peculiar to Iudaism *(e.p.* the high priest's golden plate or mitre, which S. James is Bald to have worn as president of the Church of Ierusalem), were retained. The truth seems to be that the use of it was laid aside for a time as a matter of expediency on account of the danger to the lives of the brethren, as it helped materially their persecutors to find out their hidden places of meeting. When the persecutions were over and the Church emerged from the concealment of the catacombs into the light of open day in the fourth century, then it was resumed, and this was the only reason of what could exist, for its discontinuance being of divine appointment as we have seen, though not essentially necessary any more perhaps than music, or lights or vestments.

v. 22. -Probably as a traditional memory of this recorded event, a peculiar custom exists in the East to this day. The Great Bell of the Kremlin Tower Is heard during the entire time of the chanting of the Creed. The same in other Oriental Churches, where every bell in the Church rings during the creed, as well as at the elevation and other parts of the Liturgy. *(See* O'Brien's "History of the Mass," &c.)

Again, referring to the Essenes as a religious body, the popular impression is that the Cenobite or Monastic life did not rise in Christendom till the Fourth Century. The fact is, it was coeval with, if not antecedent to, Christendom, as among the Essenes and Therapeutre, and even before them in the "school of the Prophets" in the Jewish Church. The Carthusian Monasteries in the Catholic Church give a good idea of these early monks and nuns, and their mode of life, &c. *(See* Arthur Lillies "Christianity and Buddhism," larger work). It is the natural outcome of earnestness and devotion, despised and rejected by the world that *will not* receive nor give heed to higher teachings but only to its own self-interests, insanities and follies, and reject the life of obedience to God's laws." Ye are the salt of the earth, a city set on a hill, a light shining in a dark place." Thus, with few exceptions, were these institutions of old, and still are, where the evil influence of the world has not crept in like a serpent coiling round the Tree of Life.

v. **26.** -Here we subjoin the ancient Anathema omitted in the first Edition which we are now admonished to restore to the complete Edition as now published.

Original Front Cover by Jain

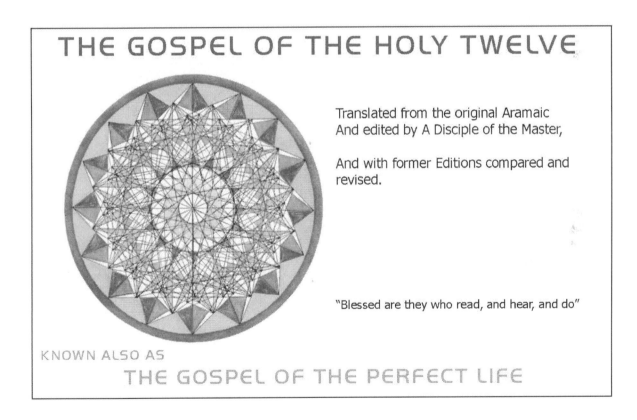

Original Back Cover by Jain

THE GOSPEL OF THE HOLY TWELVE

This rare Essene Bible was found in an urn in Tibet in a Buddhist monastery where it had been laying for 1800 years until it was recently rediscovered and translated from the ancient Aramaic (the language Jesus spoke) into English. Of the 4 translators, one of them was the famed Emmanuel Swedenborg (died 1772), the Swedish mystic and seer.

My particular interest in this manuscript is for 5 reasons:

1: I first discovered this book at the Adyar bookshop, Sydney, around the time of 1977 when I was 20 years of age and it was protruding from a top shelf. As I jumped to push the book back in place, I suddenly decided to grab it. I normally would not be standing in front of Christian material. This book changed my life.

2: As I flipped through the pages, this book had many images of Magic Squares of 3x3, 7x7 and 11x11 and sacred symbols like Pythagoras' Triangle, Egyptian Ankhs, Star of Davids, Alpha and Omega symbols, Ground Plan of the Christian Church, Labyrinth etc.

3: The Lord's Prayer, in Lection XIX, in the chapter: "Iesus Teaches Concerning Prayer" was the first book I had encountered that honored both the Father and the Mother: It begins:
"OUR FATHER-MOTHER Who Art Above And Within: Hallowed Be Thy Sacred Name In Twofold Trinity . In Wisdom, Love and Equity Thy Kingdom Come To All...". I believe that the words "In Twofold Trinity" is a clear reference to the Phi Ratio which examines a Trinity of 3 parts into a specific twofold partitioning of 2 parts, one is the larger segment and the other is the smaller segment).

4: There is the most beautiful story about a man born blind from birth who Jesus heals. It is called: "The Examination of Him Who Was Born Blind" (Lection LIV). This man becomes devoted to Iesus' teachings and asks many questions. It continues to describe the Divine Kingdom in terms of the Magic Square of 7x7 and is referred to as THE PARABLE OF THE SEVEN PALMS. From this material I have been inspired to create what I call THE THEATRE OF THE HOLY NUMBERS which is a creative re-enactment of this biblical scene. The Magic Square of 11x11, like the 7x7 is important to the Essenes as the central crosses in both also add up to the Magic Sum or Constant of the Rows and Columns. In the Brief Commentary, the Magic Square of Eleven Perfected is subtitled: "Ecclesiae Militantis Sigilum". which means the Magic Square of 11x11 is the Churches' Military Seal of Power or Defence.

5: This is a very rare book with no ISBN. It appears to be from the same body of works that documented the missing 18 years of Jesus' life deleted from the modern bibles, and is essential reading.

JAIN'S NOTES: ABOUT THE BOOK:

"THE GOSPEL OF THE HOLY 12"

Contains 193 photocopied pages, colour front and back cover, $50, originally wire bound, but now perfectly bound with spine.
Originally read and bound in the landscape format. The book has not been computerized, I have kept the original pages as they appeared in a 1974 reprinted version. Perhaps a later version will have it all computerized. Very fine print but still readable.

The front cover only says:
"Translated from the original Aramaic and edited by a Disciple of the Master..."

This is a very rare book with no ISBN; and I have not seen another in print, nor met anyone that has heard of it. It appears to be from the same body of works that documented the missing 18 years of Jesus' life deleted from the modern bibles, and is essential reading.

BACK COVER BLURB

This rare Essene Bible was found in an urn in Tibet in a Buddhist monastery where it had been laying for 1800 years until it was recently rediscovered and translated from the ancient Aramaic (the language Jesus spoke) into English. Of the 4 translators, one of them was the famed Emmanuel Swedenborg (died 1772), the Swedish mystic and seer.

My particular interest in this manuscript is for 4 reasons:

1: I first discovered this book at the Adyar bookshop, Sydney, around the time of 1977 when I was 20 years of age and it was protruding from a top shelf. As I jumped to push the book back in place, I suddenly decided to grab it. I normally would not be standing in front of Christian material. This book changed my life.

2: As I flipped through the pages, this book had many images of Magic Squares of 3x3, 7x7 and 11x11 and sacred symbols like Pythagoras' Triangle, Egyptian Ankhs, Star of Davids, Alpha and Omega symbols, Ground Plan of the Christian Church, Labyrinth etc.

3: The Lord's Prayer, in Lection XIX, in the chapter: "Jesus Teaches Concerning Prayer" was the first book to honour both the Father and the Mother: It begins:
"**OUR FATHER-MOTHER** Who Art Above And Within: Hallowed Be Thy Sacred Name In Twofold Trinity. In Wisdom, Love and Equity Thy Kingdom Come To All...".

I believe that the words "In Twofold Trinity" is a clear reference of the Phi Ratio which examines a Trinity of 3 parts into a specific twofold partitioning of 2 parts, one is the larger segment and the other is the smaller segment.

4: There is the most beautiful story about a man born blind from birth who Jesus heals. It is called: "The Examination of Him Who Was Born Blind" (Lection LIV). This man becomes devoted to Iesus' teachings and asks many questions. It continues to describe the Divine Kingdom in terms of the Magic Square of 7x7 and is referred to as THE PARABLE OF THE SEVEN PALMS. (The Yantram on the front cover is a Magic Square of 7x7 rotated 8 times, inspired from this Parable). From this material I have been inspired to create what I call THE THEATRE OF THE HOLY NUMBERS which is a creative re-enactment of this biblical scene.

The Magic Square of 11x11, like the 7x7 is important to the Essenes as the central crosses in both also add up to the Magic Sum or Constant of the Rows and Columns. In the Brief Commentary, the Magic Square of Eleven Perfected is subtitled: "Ecclesiae Militantis Sigilum". which means the magic Square of 11x11 is the Churches' Military Seal of Power or Defense.

The GOSPEL OF The HOLY TWELVE
aka The GOSPEL OF The PERFECT PEACE

This rare Essene Bible was found in an urn in Tibet in a Buddhist monastery where it had been laying for 1800 years until it was recently rediscovered and translated from the ancient Aramaic (the language Jesus spoke) into English. Of the 4 translators, one of them was the famed Emmanuel Swedenborg (died 1772), the Swedish mystic and seer.

My particular interest in this manuscript is for 5 reasons:

- I first discovered this book at the Adyar bookshop, Sydney, around the time of 1977 when I was 20 years of age and it was protruding from a top shelf. As I jumped to push the book back in place, I suddenly decided to grab it. I normally would not be standing in front of Christian material. This book changed my life.

- As I flipped through the pages, this book had many images of Magic Squares of 3x3, 7x7 and 11x11 and sacred symbols like Pythagoras' Triangle, Egyptian Ankhs, Star of Davids, Alpha and Omega symbols, Ground Plan of the Christian Church, Labyrinth etc.

- The Lord's Prayer, in Lection XIX, in the chapter: "Jesus Teaches Concerning Prayer" was the first book to honour both the Father and the Mother: It begins:
"OUR FATHER-MOTHER Who Art Above And Within: Hallowed Be Thy Sacred Name In Twofold Trinity. In Wisdom, Love and Equity Thy Kingdom Come To All...".
I believe that the words "In Twofold Trinity" is a clear reference of the Phi Ratio which examines a Trinity of 3 parts into a specific twofold partitioning of 2 parts, one is the larger segment and the other is the smaller segment.

- There is the most beautiful story about a man born blind from birth who Jesus heals. It is called: "The Examination of Him Who Was Born Blind" (Lection LIV). This man becomes devoted to Iesus' teachings and asks many questions. It continues to describe the Divine Kingdom in terms of the Magic Square of 7x7 and is referred to as The PARABLE Of The SEVEN PALMS. From this material I have been inspired to create what I call The THEATRE Of The HOLY NUMBERS which is a creative re-enactment of this biblical scene. The front cover design of this book is the Magic Square of 7x7 Yantram rotated upon itself 8 times at 22.5°. The Magic Sum of its columns, rows and diagonals adds up to 175.

- The Magic Square of 11x11, like the 7x7 is important to the Essenes as the Central Crosses in both also add up to the Magic Sum or Constant of the Rows and Columns. In the Brief Commentary, the Magic Square of Eleven Perfected is subtitled: "Ecclesiae Militantis Sigilum" which means the Magic Square of 11x11 is the Churches' Military Seal of Power or Divine Defence.

Made in United States
Troutdale, OR
07/19/2024

21415761R00108